NUCLEIC ACIDS - FROM BASIC ASPECTS TO LABORATORY TOOLS

Edited by **Marcelo L. Larramendy**
and **Sonia Soloneski**

Nucleic Acids - From Basic Aspects to Laboratory Tools
http://dx.doi.org/10.5772/60691
Edited by Marcelo L. Larramendy and Sonia Soloneski

Contributors

Márcia Aparecida Sperança, Aline Diniz Cabral, Rodrigo Buzinaro Suzuki, Andreia Moreira Dos Santos Carmo, Ivo Sirakov, Gisele Gouveia, Suzete Ferreira, Sheila Siqueira, Juliana Pereira, Lifang Yang, Zhijie Xu, Theam Soon Lim, Qiuting Loh, Luis M. Alvarez-Salas, Ana Gabriela Leija-Montoya, María Luisa Benítez-Hess, Erin R. Murphy, Yahan Wei, Gabrielle Heilek, Haifeng Chen, Joao Carlos Oliveira Guerra

Published by InTech

Janeza Trdine 9, 51000 Rijeka, Croatia

Publishing Process Manager Iva Simcic

Technical Editor InTech DTP team

Cover InTech Design team

First published March, 2016

Printed in Croatia

Additional hard copies can be obtained from orders@intechopen.com

Nucleic Acids - From Basic Aspects to Laboratory Tools, Edited by Marcelo L. Larramendy and Sonia Soloneski

p. cm.

ISBN 978-953-51-2264-7

For EU product safety concerns:
IN TECH d.o.o., Prolaz Marije Krucifikse Kozulić 3, 51000 Rijeka, Croatia,
info@intechopen.com or visit our website at intechopen.com.

Meet the editors

Marcelo L. Larramendy, Ph.D., serves as Professor of Molecular Cell Biology at the School of Natural Sciences and Museum (National University of La Plata, Argentina). He is appointed as Senior Researcher of the National Scientific and Technological Research Council of Argentina. He is a former member of the Executive Committee of the Latin American Association of Environmental Mutagenesis, Teratogenesis and Carcinogenesis. He is author of more than 470 contributions, including scientific publications, research communications and conferences worldwide. He is a recipient of several national and international awards. Prof. Larramendy is a regular Lecturer at the international A. Hollaender Courses organized by the IAEMS and former guest scientist at the NIH, the United States, and University of Helsinki, Finland. He is expert in Genetic Toxicology and is, or has been, referee for more than 30 international scientific journals. He is also a member of the International Panel of Experts at the International Agency for Research on Cancer (IARC, WHO, Lyon, France) in 2015 for the evaluation of DDT, 2,4-D and Lindane. Presently, Prof. Dr. Larramendy is the Head of the Laboratory of Molecular Cytogenetics and Genotoxicology at the UNLP.

Sonia Soloneski is PhD in Natural Sciences and Professor Assistant of Molecular Cell Biology at the Faculty of Natural Sciences and Museum of La Plata, National University of La Plata, Argentina. She became a member of the National Scientific and Technological Research Council (CONICET) of Argentina in Genetic Toxicology field. Presently, she is member of the Latin American Association of Environmental Mutagenesis, Teratogenesis and Carcinogenesis (ALAMCTA), the Argentinean Society of Toxicology (ATA), and the Argentinean Society of Genetics (SAG). She is author of more than 290 scientific publications on the field, including scientific publications in research papers, reviewed journals, and conferences worldwide. She is a referent for on issues related to fields of Genetic Toxicology, Mutagenesis, and Ecotoxicology.

Contents

Preface

Nucleic acids, the most common of which are DNA and RNA, have a wide variety of functions in the main system of life, expression, maintenance, and so on. In the postgenomic era, when the primary structure of our genome has been fully determined, though not yet fully understood, a significant challenge for investigation is to determine not only the role of several small molecules of nucleic acids with promising functions but also the modification and employment of novel nucleoside analogues developed for biological research and medical applications. Nucleic acids are commonly used as laboratory tools for modulating gene expression and have the potential to be widely used in the fields of medicine and therapeutics. They can recognize and exhibit a much stronger affinity with their corresponding counterparts, such as DNA or RNA, to form a thermodynamically stable variable molecular structure and thus can be used in gene therapy to silence genes.

This volume contains a series of chapters that highlight the development and status of the various aspects of the nucleic acids related to DNA chemistry and biology and the molecular application of these small DNA molecules and related synthetic analogues within biological systems.

This book, *Nucleic Acids,* is organized into 11 chapters containing methods in DNA extraction, molecular diagnosis, gene-silencing therapeutics, and bacterial gene regulation by RNA thermometer. The first chapter is a review presentation describing physical properties of DNA to propose a new model for allowing different DNA nucleation strategies. The second chapter describes the fundamental principles of different methods for nucleic acid sample preparation and extraction, such as column-based methods using silica membranes and traditional ones without a column purification procedure; a comparative analysis of the use of these methods in DNA and RNA extraction from various biological and clinical samples; and the application of nucleic acids in molecular diagnosis. The third chapter describes the factors affecting the formalin-fixed and paraffin-embedded tissue nucleic acid extraction process, compares the available protocols, and describes the modifications of some protocols. The fourth chapter provides a summary of the application of the molecular methods for the detection of two important foodborne viruses. The fifth chapter is a an overview of the contribution of the nucleic acid discovery to the diagnosis and molecular epidemiology of pathogenic microorganisms and its relevance for veterinary and human health, principally after the introduction of polymerase chain reaction (PCR) and post-PCR techniques employed in infectious diseases. The sixth chapter of this book provides a description of the use of nucleic acids in molecular diagnostics, focusing on the impact of molecular testing for the management of infectious diseases, emphasizing personalized healthcare and medical needs in the developing market for molecular testing. The seventh chapter provides a review of nucleic acid aptamers for viral detection and inhibition, focusing on the most successful antiviral aptamers reported and including a description of some of the novel methods developed for their use as diagnostic and therapeutic tools. The eighth chapter highlights the current advantages and flexibilities of the DNA G-quadruplex and its potentiality in conventional diagnostic platforms that use protein enzymes. The ninth chapter includes advances in medical aspects related to gene-silencing therapeutics, focusing mainly on DNAzymes and siRNA and their integral role in future antigene therapies. Finally, the two last chapters summarize the current state of knowledge on bacterial RNA thermometers, including several aspects and techniques applied to identifying different families of RNA thermometers and a review on the thermodynamics of the denaturation transition of DNA duplex oligomers in the context of nearest neighbor models.

It is our sincere hope that this book on some widely used methods for studying the several types of nucleic acids and their related molecules will inspire future translational studies focusing on targeting nucleic acid regulatory mechanisms as a strategy in several therapies in a variety of illness and diseases.

This book will serve as a textbook for the research community, academia, and the industry and provides up-to-date state-of-the-art information presented by recognized international experts in the field of the molecular biology. They have contributed to the publication of this book of high importance to researchers, scientists, engineers, and graduate students who make use of these different investigations to understand the hazardous implications in the use of nucleic acids. The provision of the publishing Web platform by the publisher InTech is gratefully acknowledged.

Prof. Dr. Marcelo Larramendy
Principal Researcher CONICET
School of Natural Sciences and Museum
National University of La Plata
La Plata, Argentina

Dr. Sonia Soloneski
School of Natural Sciences and Museum
National University of La Plata
Argentina

Chapter 1

Nucleic Acid Isolation and Downstream Applications

Ivo Nikolaev Sirakov

Additional information is available at the end of the chapter

http://dx.doi.org/10.5772/61833

Abstract

Nucleic acids are not only a source of life but also a means of observing, understanding, and regulating it. Nucleic acids, DNA and RNA, and their characteristics are discussed in other chapters of the book. This chapter describes the fundamental principles of different methods for nucleic acid sample preparation / nucleic acid extraction, such as column-based methods using silica membranes and traditional ones without a column purification procedure (commercially available or homemade). Other topics discussed here include comparative analysis of the use of these methods in DNA and RNA extraction from a variety of biological and clinical samples, as well as the relationship between the type of sample, the method used and the quality and amount of extracted DNA or RNA. Finally, the chapter outlines the application of nucleic acids in the diagnosis of various diseases, in scientific research, and bird sex determination by downstream applications such as restriction enzyme analysis, polymerase chain reactions (PCR, reverse transcription-PCR, real-time PCR), and different sequencing methods (Sanger, cycling sequencing, and next-generation sequencing).

Keywords: DNA extraction, RNA extraction, PCR methods, next-generation sequencing

1. Introduction

1.1. Nucleic acid isolation and downstream applications

The specific properties of nucleic acids have been widely employed in the development of different molecular methods and mathematical models for their analysis. These methods are applied to identify microorganisms and genetic predispositions, to detect different mutations and determine their role in antibiotic resistance, to study phylogenetic relationships, and so on. What all these methods share in common is their starting point: obtaining a purified nucleic acid sample.

2. Nucleic acid extraction methods

Since nucleic acid extraction is a starting point in a vast array of downstream applications, the high quality of nucleic acids in the starting samples is a key factor for the success of the subsequent steps of analysis. Thus, nucleic acid extraction could be defined as a series of steps to obtain nucleic acid samples/materials of particular purity that are free of impurities and are suitable for different downstream application steps. The purpose of nucleic acid extraction methods is to disintegrate the cell envelope and achieve maximum elimination of lipids and proteins to obtain pure DNA and/or RNA. This is principally based on heat adsorption on silica membranes/beads, anion exchange chromatography, sedimentation/precipitation, and use of magnetic particles. These methods yield initial nucleic acid samples of different purity and concentration depending on the original sample (bacteria, viruses, tissues).

The choice of method – in view of optimal time/quality balance – depends on the aim of the study, the type of analysis, the type of nucleic acid, and the cost. It is important to provide appropriate conditions for nucleic acid extraction in order to avoid nucleic acid degradation due to oxidation by reactive oxygen species generated during respiration in vivo or, extracellularly, by mechanisms involving metal ions [1–4]. Nucleic acid degradation can result from hydrolysis of the 3′-5′ phosphodiester bonds catalyzed by metal complexes as well as from the spontaneous breakage of these bonds due to transesterification via a nucleophilic attack at the phosphorus atom by an adjacent 2′-hydroxyl group [1].

What accounts for the differences between the methods for extraction of DNA and RNA is their different stability. RNA includes ribosomal RNA (rRNA) 80%, mitochondrial RNA (mtRNA), messenger RNA (polyadenylated – poly A^+ in eukaryotic cells) (mRNA) 1–5%, transfer RNA (tRNA), and microRNA molecules (miRNA). There may be different amounts of mRNA in cells: from large quantities to just five copies per cell. In fact, mRNA is the RNA of choice in reverse transcription and cDNA synthesis. RNA molecules are susceptible to degradation by ozone in the air; ozone is highly reactive regardless of whether the RNA sample is liquid or solid [5]. Another factor that plays a role in RNA degradation is water, as it makes the transfer of protons possible and serves as a source of hydronium or hydroxyl ions. That is why dehydration has a protective effect against RNA degradation [6]. Nucleic acids, and especially RNA molecules, are also sensitive to nucleases. Therefore, in RNA extraction procedures, it is essential to ensure an RNase-free fraction and a means to quickly cool down the sample. This illustrates the importance of providing all necessary work facilities: use of BSL-1 or 2 laminar flow cabinets (depending on the type of biological material) is recommended for nucleic acid extraction; ultraviolet (UV) germicidal irradiation of both the premises and the laminar flow cabinets should be done, with irradiation of the premises done the night before (irradiation immediately before work may lead to degradation due to residual UV light). The same principle applies to UV germicidal irradiation of plastic labware, e.g., microtubes, pipette tips, etc. (if not commercially sterile, DNase- and RNase-free). Another important detail is that talc-free gloves should be used because talc may inhibit some downstream analyses such as PCR, reverse transcription, and real-time-PCR.

The quality of the extracted nucleic acids also depends on the quality of the starting sample. In fact, all manufacturers of nucleic acid isolation kits recommend that fresh starting material be used. If this is not possible – as is often the case with diagnostic samples – they can be stored for 24–48 h at 4°C, or for longer periods of time at –80°C or in liquid nitrogen, preferably using protective buffers, especially for samples intended for RNA analysis. There are paper matrices especially developed for storage and transport of blood samples at room temperature – dried blood spot sampling. In this train of thought, it has to be kept in mind that heparin, which is used as an anticoagulant, may inhibit some PCR reactions [7] and should, therefore, be avoided or removed. In the case of clotted blood samples, the coagulum can be treated as an organ sample. The spleen and the liver are transcriptionally active organs and they have a very high RNA content. Because of that, if the samples are intended for DNA analysis, they have to be treated with RNase prior to column purification. When the aim is RNA analysis, it is particularly important to protect the RNA against degradation, else the low-frequency transcripts could be lost and would not be detected in the downstream steps of analysis. Moreover, in microarray analysis, degraded RNA molecules may fail to successfully bind to the complementary site due to loss of the complementary sequence. That is why, in RNA extraction, frozen samples should be mechanically processed (homogenized) prior to thawing, and fresh tissue samples should be ground in liquid nitrogen or by other means of cooling. This is not as essential in DNA extraction, since DNA molecules are relatively more robust, but is recommended.

Another key step is cell lysis, which – if incomplete – would result in reduced yield and column blocking and, in turn, in lower purity. There are different ways to aid the process of cell lysis: in the case of cell cultures or bacterial cultures, depending on the aim of analysis, they can be washed in PBS (phosphate buffered saline) or physiological saline and resuspended in ddH_2O and/or subjected to several freeze/thaw cycles. In the case of mucous samples (nasal discharges, sputum, intestinal loops), it is good to first decrease the viscosity of the material (using a mucolytic –"mucus-dissolving" agent, e.g., acetylcysteine). The mechanical processing of samples from insects, plants, feces, organs requires 50–200 mg of sample in most kits. Gram-positive bacteria are treated with lysozyme; yeasts, with zymolyase or lyticase; and paraffin-embedded tissues are treated with xylene to remove the paraffin. Other approaches that can be applied to disrupt cell envelopes include: osmotic shock, which is suitable for Gram-negative bacteria, cell cultures and erythrocytes; chaotropic salts, for all types of samples with the exception of some Gram-negative bacteria, owing to the greater thickness of their peptidoglycan layer; enzymatic degradation (lysozyme, proteinase K), which is often combined with osmotic shock or freeze/thaw cycles (for DNA extraction from hair, feathers etc.); and detergents, for tissue cultures. Should a sample remain not fully lysed, one way to overcome the problem is to centrifuge the mixture and use the supernatant prior to column loading or alcohol supplementation, in case the protocol includes such a step.

Sometimes the samples may be "old", i.e., stored for a long time at –20°C. In such samples, the chemical bonds in the DNA molecules may have become weaker. Then, in the vortexing steps in the column-based and solution-based methods, the DNA becomes, more often than not, degraded, resulting in smeared DNA bands (Figure 1). To avoid this, such samples should be kept frozen during the process of mechanical homogenization, and instead of vortexed, should be gently homogenized by slowly turning the microtubes upside down and back several times to adequately mix the reagents and, to a large extent, preserve the intactness of the DNA molecules.

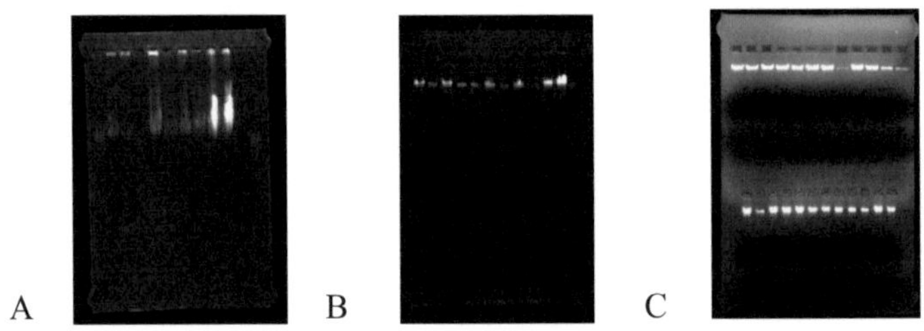

Figure 1. DNA from blood samples. Smeared DNA bands (A), DNA with double-strand breaks (B) and normal intact DNA bands (C). First Report and Final Report of a research grant awarded to Dr. Ivo Sirakov, 2007–2008 – Medicine and Biotechnology – Aids and infectious diseases, Transmissible Spongiform Encephalopathies, funded by the World Federation of Scientists, Geneva, Switzerland.

The methods for preparation of nucleic acid samples can be grouped into thermal extraction, solution-based methods (homemade and commercial kits), column-based methods, and ones that use magnetic particles.

Thermal extraction is a quick and low-cost method that does not require special reagents. It can be used to extract DNA from pure bacterial cultures. There are different variations of the method [8, 9], but they generally include the following procedure: cultures grown 18–24 h are used; dilutions are prepared in Ultra-pure 18.2 MΩ DNase/RNase-free water (when using culture broth, 1:40, or when using agar cultures, 0.5 McF in 1 mL); the samples are heated at 100°C for 5–15 min and then centrifuged; the supernatant is taken and stored at –20°C.

Despite its advantages, this method has some limitations: it cannot eliminate low molecular weight peptides and gives a low 260/280 nm ratio (purity), i.e., about 0.600. This limits the use of the extracted DNA only to conventional PCR.

Thermal extraction is part of the official VTEC *E. coli* diagnostic procedure, which is based on detection of *eae* gene fragments up to 384 bp in length (15). This method, however, appears inapplicable to amplification of larger fragments, especially the *E. coli* 16s RNA gene, which is 1,465 bp in size [10].

The next group of methods, the solution-based ones, in principle, includes the following basic steps: lysis, RNase treatment (if applicable), protein precipitation (two fractions are formed), separation of the fraction that contains the nucleic acids; their precipitation, washing, drying, and regeneration.

One of the earliest and most common solution-based methods is phenol–chloroform extraction. It also has various modifications. Volkin and Carter [11] first developed a method for

RNA extraction with 2 M guanidine hydrochloride, chlorophorm, and alcohol, in which guanidine acts as a deproteination agent and protects the RNA molecules by denaturing the proteins and RNases [12]. After a series of different steps, there is an extraction step using guanidinium thiocyanate–phenol–chloroform [13]. The phenol–chloroform combination (in a 1:1 ratio) gives better protein denaturation (by forming two fractions following centrifugation: a bottom organic phase and an upper aqueous phase) and reduces the amount of poly(A)+ mRNA in the organic phase as well as of insoluble RNA–protein complexes in an intermediate phase [14]. What is more, chloroform prevents the retention of water in the aqueous phase (water can degrade RNA molecules; see above), which results in higher yield [15]. To avoid foam formation, isoamyl alcohol can be added (chlorophorm–isoamyl alcohol, 24:1). It is the acidic properties of phenol that actually determine the partitioning of DNA and RNA in a separate phase: at neutral and slightly alkaline pH (i.e., pH 7–8), DNA and RNA remain in the aqueous phase, since the phosphate diesters are negatively charged. At lower pH (optimal pH 4.8), DNA partitions in the bottom phase, whereas RNA remains in the aqueous phase. This is due to the fact that the phosphate groups in DNA are more prone to neutralization compared to those in RNA [16, 17]. This method is commonly used for DNA extraction from liquid samples, although it also gives good results with tissue cultures, cell cultures etc., provided that they are first homogenized and disintegrated using a lysis buffer (commercial kit or homemade, for example – 10 mM Tris, 1 mM EDTA (ethylenediaminetetraacetic acid) and 0.1 M NaCl) + 20–50 μL of 10–20 mg/mL proteinase K and 1–18 h of incubation at 50–60°C.

There are some specifics in RNA extraction procedures: to protect RNA molecules against RNase attack, chaotropic salts are added to extraction buffers. High-purity RNA can be obtained by guanidine treatment followed by gradient ultracentrifugation in CsCl [12], cesium trifluoroacetate, or LiCl. A disadvantage is the need to use an ultracentrifuge and the fact that the method is time-consuming (it takes about 16 h).

In the case of notoriously difficult samples, Birnboim [18] recommend the use of a combination of SDS (sodium dodecyl sulfate) and urea to more effectively inhibit leukocyte RNases.

During extraction, beta-mercaptoethanol can be added to denaturate RNases by reducing disulfide bonds and to aid the release of RNA from RNA–protein complexes.

For RNA extraction, phenol and guanidinium isothiocyanate mixtures (Trizol reagent) are available under different commercial brand names and chlorophorm is then added in the extraction process to separate the phases [19].

As a whole, solution-based methods, and especially their homemade versions, have the disadvantage that they use toxic reagents and, therefore, a fume hood is a must; moreover, precision in phase separation is difficult to achieve; the extracted nucleic acid samples are of lower purity (especially in homemade methods) compared to those obtained by column-based methods; the obtained samples have limited use – mainly for conventional PCR. Commercial kits and protocols, albeit validated for downstream applications, do not solve the problem of reagent toxicity and subjective factors related to the person who performs the procedure.

The main advantage of these methods is that they allow for a proportionally larger initial material to be used.

Column-based methods generally use the fact that DNA and RNA molecules are negatively charged to capture them by silica membrane and ion exchange chromatography. These methods include lysis in one or two steps (in silica membranes – chaotropes and proteolytic lysis, and in anion exchange – detergent and enzymatic digestion); then, loading the liquid sample onto a column and centrifugation for the purpose of nucleic acid binding in a high-salt environment; treatment of the column with lysis buffer and/or directly with wash buffer (containing alcohol and high salt); free centrifugation to eliminate the remaining buffer (ethanol*); elution with ultra-pure RNase/DNase-free H_2O (≈pH 8.0) or low-ionic-strength buffer in the case of silica membrane and higher-salt buffer (not suitable for most downstream applications) in the case of anion exchange; incubation of room temperature for 1–3 min and centrifugation. It is possible to repeat the washing and elution steps to achieve further purification (if the column is treated with DNase or RNase after the first elution) or to elute a higher amount of nucleic acids. The standard procedure takes about 20 min per sample or 35–40 min in the case of DNase or RNase treatment. It may be possible to use more starting sample than recommended but it should be kept in mind that this could overload the purification column, resulting in lower yield and/or higher percentage of impurities.* Ethanol remaining in the end-product may lead to the escape of the nucleic acid from the wells during gel electrophoresis and may block restriction analysis, PCR, and other enzyme reactions.

To directly extract mRNA, without an initial procedure of total RNA extraction, oligo(dT) affinity chromatography is used. In this method, oligo(dT) is bound to cellulose or paramagnetic particles. The essence of the method lies in binding of mRNA poly(A^+) tails to the oligo(dT) fragments attached to the matrix in high-salt conditions. Then, high-salt wash buffer is applied, followed by low-salt wash buffer, and elution under low ionic strength. This method, however, is only limited to eukaryotic cells, since their mRNA is polyadenylated. Another drawback is that the samples may be selectively enriched in mRNAs with shorter poly(A) tracts. What should be avoided is using a maximum amount of sample and overloading, as indicated by higher viscosity and a mucus-like look of the medium.

Another approach to nucleic acid extraction is based on magnetic particles, i.e., the so-called charge-based method: at pH ≤ 6.5, the surface of magnetic particle is positively charged and binds nucleic acids. Depending on the purpose, the unwanted nucleic acid can be eliminated by DNase or RNase treatment. Then, the nucleic acid molecules remaining bound to the particles are released by changing the pH: at pH ≥ 8.5, the surface is neutral and the nucleic acid molecules are released.

In the case of DNA extraction, the comparative analysis of all these methods shows that the quantity/quality of extracted DNA is inversely proportional to the extraction time.

It is commonly accepted to assess the purity of extracted nucleic acid samples based on the ratio of nucleic acids to proteins (impurities). This ratio is determined by measuring the absorbance of the samples at 260 and 280 nm and calculating the 260/280 nm ratio (the absorbance at 260 nm reflects the mean absorbance of purines and pyrimidines [20, 21]. In the case of DNA samples, it is recommended to make a background correction at 320 nm, which accounts for turbidity. The sample purity reflects the degree to which different contaminants have been eliminated in the nucleic acid extraction procedure. There are different possible sources of contamination. These include the reagents – salts and residual buffer (especially the alcohol-containing wash buffer), as well as different compounds present in the starting material – polysaccharides, phenolic compounds, and DNA or RNA and proteins – nucleases. Elimination of these contaminants is an intrinsic part of the procedures in all the nucleic acid extraction methods. Moreover, there may be included additional purification steps: ultracentrifugation to eliminate high molecular weight polysaccharides; use of beta-mercaptoethanol, dithiothreitol, sulfite, etc.; RNase or DNase can be added and care should be taken to protect the samples from exogenous RNases or DNases of different nature (the samples themselves, human skin, or the laboratory environment), by following the principles of Good Laboratory Practice. It is also recommended to use DEPC, which inactivates RNases in solutions, with the exception of solutions that contain primary amines, such as Hepes buffer and Tris, as they reduce its effect. It is noteworthy that autoclaving will not destroy RNase activity. In plant samples, contamination is often due to polysaccharides and polyphenols, which are eliminated by using polyvinylpyrrolidone. DNA and RNA are also considered contaminants: in DNA analysis, RNA acts as a contaminant and vice versa. Hence, DNase or RNase is used to eliminate contaminating DNA or RNA, respectively. This is so, first of all, because both DNA and RNA contribute to the total nucleic acid content, which – if too high – may block downstream PCR. On the other hand, in gene expression analysis, the methods are sensitive to DNA contamination. For example, reverse transcription (RT) and microarray methods require high RNA purity, since small DNA fragments may anneal to the primers, giving a false positive result, whereas other contaminants (phenol, ethanol, and salts in RT) may react with enzymes, blocking the reaction or increasing the background signal. Conversely, methods such as Northern blotting are not as sensitive to contamination. Thus, sample purity may vary in a certain range – from 0.600 to over 3.0 (mostly for RNA) – depending on the aim of analysis, the extraction method used, the starting material, and the operator. Samples are considered to be of good quality if their 260/280 nm ratio is over 1.7, which is satisfactory for most downstream applications.

However, it is not only the quality of the extracted nucleic acids that is considered important but also their quantity. Depending on the downstream application, if the concentration of nucleic acid in a sample is low, this may, at least in part, be compensated for by using a greater volume of sample in the reaction mixture (the sample purity should be considered as well). For example, if there is insufficient concentration of nucleic acid, non-specific products may be amplified in PCR; or some expected fragments may appear missing in restriction enzyme analysis; or short read lengths may be generated in sequencing. Nucleic acid concentration that is too high may also have adverse effects: amplification of non-specific fragments in PCR or lack of product due to reaction blocking; retention of nucleic acid in gel wells during electrophoresis; incomplete digestion in restriction enzyme analysis (which may be compensated for by adding a proportionally higher amount of enzyme and/or extending the incubation time); or high background in sequencing procedures.

To concentrate, purify DNA and reduce the salt content in DNA samples, precipitation is used (sample: 99% molecular grade ethanol (or isopropanol) 1:1 + 2 to 5% 3.5–7 M ammonium acetate) with 30–60 min incubation at –20°C, centrifugation at 4°C at maximum speed, washing in maximum volume of 70% ice-cold ethanol (–20°C), centrifugation, drying for 3–7 min and resuspending in a desired volume of ultra-pure water, ¼ TE buffer or 1× TE buffer, depending on the downstream application and the expected duration of storage.

Another way to determine the quality and quantity of extracted nucleic acids, apart from spectrophotometric analysis, is by gel electrophoresis (GE), which is informative of fragmentation and presence of impurities. Although, in some cases, GE may be sufficient when the researcher is experienced, it is still recommended to use both methods together.

The type of storage and its duration are crucial for the downstream applications. In the case of DNA samples, both the temperature storage and the buffer composition are important factors. Storage at –20°C in 1× PCR buffer for 100 days gives very good results [22]. There are reports that DNA stability can be enhanced by adding 50% glycerol, which limits the formation of ice crystals [23]. Overall, storage at –20°C in commercially available elution buffers (EB) gives stable DNA for use even after a year of storage (EDTA as a component of EB protects DNA against degradation), provided that repeated freezing and thawing are avoided. RNA samples are stored at –80°C in stabilizing buffers that contain EDTA; even storage microcapsules have been developed [1].

3. Nucleic acid and Restriction Enzyme Analysis (REA)

Restriction analysis is an easy-to-perform, inexpensive, and relatively fast method for the study of point mutations and identification of methylated regions in DNA. It can also be used for restriction profiling of micro- and macroorganisms and can serve as a basis for phylogenetic analysis.

In the case of fragmented nucleic acids, e.g., viruses with a segmented genome (Rotavirus), direct fingerprinting is applied, in which individual nucleic acid segments – due to different mobility in an electric field – are distributed at a different distance in an agarose or polyacrylamide gel [24].

Direct electrophoresis, however, does not work in the case of non-segmented nucleic acids. That is why restriction enzymes (restrictases) are used. Restriction enzymes cut the nucleic acid molecule at a specific nucleotide sequence that they recognize. The method is both applicable to total homogeneous (from a single species) DNA and to PCR amplification products. The requirements for the quality of the nucleic acid sample are laid out above (see nucleic acid extraction methods). Additionally, it is recommended to purify the PCR amplifi-

http://dx.doi.org/10.5772/61833

cation product in a gel (most kits recommend 2% gel, although 1.5% gels give better purification) or directly by a column, before restriction. Thus, if there is not enough DNA product in the reaction mixture, more DNA template sample can be added instead of water. It is essential to keep the enzyme/buffer/reaction volume ratio specified in the instructions provided by the enzyme manufacturer.

4. Nucleic acid detection

Detection of nucleic acids and/or traces of them is widely applied in various areas (e.g., biodiversity assessment, marker-assisted selection, molecular diagnostics of infectious diseases, and genetic disorders, etc.), as well as in a range of other fields of industrial and social importance, e.g., food and pharmaceutical industry, healthcare, forensics, etc., only to name a few. For the purpose of nucleic acid detection, there have been developed a number of methods based on hybridization (such as in situ hybridization, molecular beacon) and polymerase chain reactions (PCR, reverse transcription-PCR, real-time PCR).

Nucleic acid hybridization is based on the ability of two complementary nucleic acid strands, at specific conditions, to form a stable double helix. This is mediated by purine–pyrimidine base pairing through hydrogen bonds as first described by Watson and Crick [25]. When hybridization is employed for experimental purposes, a synthetic nucleic acid fragment, the so-called probe, is prepared such that it is labeled (tagged) with a molecule that is easy to detect (the so-called reporter). Reporter molecules were initially radioisotopes, until, in 1981, Langer et al. [26] introduced non-isotopic labeling methods using avidin–biotin binding (covalently bound to the C-5 position of the pyrimidine ring), fluorescent or chemiluminescent dyes. There are now commercial ready-to-use probes for specific diagnostic purposes.

What marked a real turning point in molecular biology was the development of polymerase chain reaction (PCR) by Saiki et al. [27], which basically includes direct in vitro synthesis and multiplication (amplification) of a specific target DNA sequence enclosed between two synthetic oligonucleotides.

Various modifications and versions of PCR have been developed since. This part of the chapter discusses the basic principles underlying conventional PCR, real-time PCR, and reverse-transcription PCR. The mechanisms of different PCRs are illustrated in Figure 2 (Since the structure of nucleic acids is described in greater detail in other chapters, it is only roughly sketched here to show the underlying principles of the reactions).

In order to design an efficient and cost-effective PCR procedure, it is essential to properly choose the reaction components and their precise concentrations: Taq DNA polymerase, buffers, deoxynucleoside triphosphates (dNTPs), $MgCl_2$, DNA template, and oligonucleotide primers [28, 29]. It is the primers [30] and Taq DNA polymerase [31] that are considered the most important factors that determine the sensitivity and effectiveness of the protocol. Another key factor is the manufacturer, accounting for different formulations, assay conditions, and/or unit definitions [32].

A standard PCR mixture should include the following main components:

Template DNA – its quality requirements are described in the "**Nucleic acid extraction methods**" part; *Primers* – specific or random complementary oligonucleotides of a different length: a forward primer for the 3′–5′ DNA strand and a reverse primer for the 5′–3′ strand. The primers are particularly important for the reaction sensitivity [30]. That is why, if you are planning to use primers reported by other authors, it is essential to first check their sequences for complementarity and completeness. (It is more often than not that erroneous primer sequences may be published, even in some prestigious journals.) Another point to consider when designing the primer sequences is the GC content, which should be about 50%; the two primers should also have similar melting temperature (*Tm*) and should not be complementary to one another but only to the target sequence, which should be conservative. In some cases, there may be differences in the sequences targeted by the primers due to mutations (in the genomes of viruses and bacteria). It is then recommended to consider all possible combinations of primer sequences, using the nucleotide coding system for mixed bases, e.g., K (G or T), Y (C or T), etc. **Such differences in the sequences targeted by the primers are used for detection of single-nucleotide polymorphisms (SNP) by real-time PCR** (see below). The primer stability depends on the degree of complementarity and the type of bonds at both ends: for example, the 3′ end should be unstable (to aid the polymerase activity), and the 5′ end should be stable [33], i.e., should contain G or C in the last three bases at the 5′ end. That is why, in some cases, a single-base mismatch in the 3′ end of the primer may not be a problem.

Random primers (RP) – These are short, synthetic, single-stranded DNA segments that are 6 (hexamers) to 10 (decamers) nucleotides in length. They consist of every possible combination of bases. In other words, in the case of hexamer primers, there must be 4^6 = 4,096 different combinations. Because of that, RP can anneal to any section of the nucleic acid template. The RP approach was described in the late 20th century and is both applicable to analysis of RNA [34, 35] and DNA [36]. The technique based on RP later evolved into RAPD–PCR (random amplified polymorphic DNA), which is a powerful typing method for bacterial species and is also commonly used in construction of genetic maps and fingerprinting libraries and identification of molecular markers [37–39]. (For details see the cited references.)

Taq polymerase is a DNA polymerase from the bacterium *Thermus aquaticus*. It has served as a basis for development of different polymerase enzymes: long range, which allows for incorporation of nucleotides up to 5–10 kb; high fidelity, which includes proofreading exonuclease activity capable of repairing mismatches introduced during strand elongation. The choice of polymerase depends on the method and downstream applications: multiplex PCR, colony PCR, low-copy PCR assay, for difficult (GC-rich) templates, cloning, library preparation, genotyping, etc.

Buffer system (Tris-HCl, $(NH4)_2SO_4$, K/NaCl, $MgSO_4$) including *deoxynucleoside triphosphates* (dNTPs – dATPs, dGTPs dCTP, and dTTPs), *$MgCl_2$* and 18.2 MΩ DNase/RNase-free *H_2O* – It

http://dx.doi.org/10.5772/61833

serves as the PCR medium. $MgCl_2$, and particularly Mg^{++}, plays a role in the elongation step as a polymerase cofactor. Additionally, Mg^{++}, along with other cations present in the mixture, reacts with the negatively charged dNTPs (four oxygen atoms surrounding the phosphorus atom (Fig 2) and DNA [40–41]. High salt concentration will lead to non-complementary annealing of DNA strands or to an increase in the DNA denaturation temperature. The buffer also plays a role in maintaining a stable pH in the reaction mixture. PCR products are identified by gel electrophoresis in 1× TBE or 1×TAE buffer and 1.5–3.0% agarose gel. To visualize the results, ethidium bromide is added to the gel (at a concentration of 1 μg/mL). Ethidium bromide binds DNA non-specifically, which allows the DNA fragments to be visualized by UV illumination. Other dyes that non-specifically bind to DNA, e.g., SYBR Green and others, can also be used.

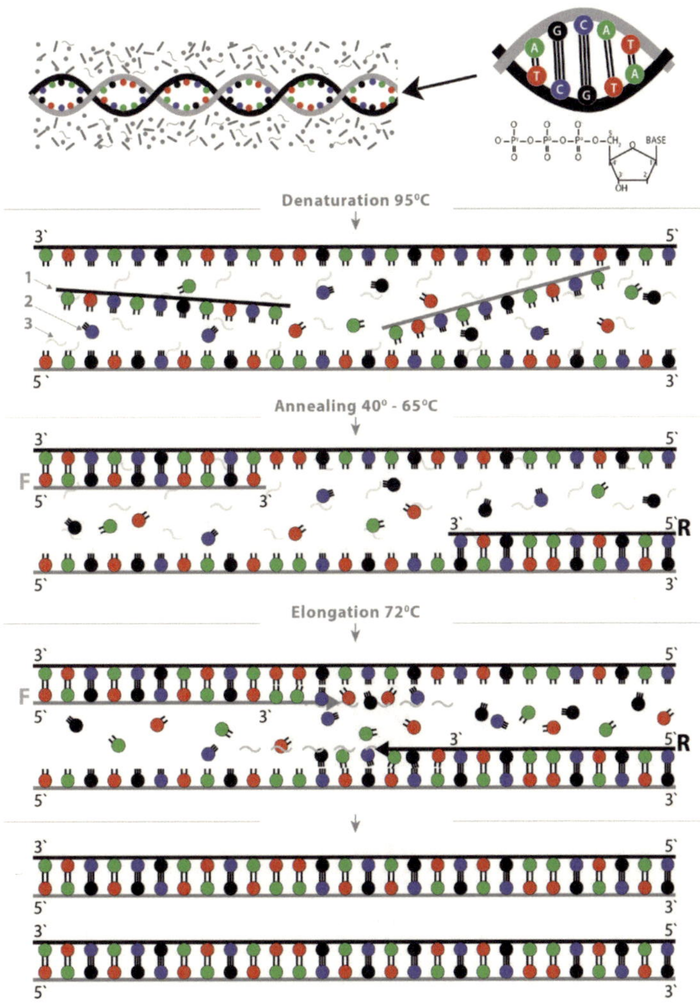

Denaturation
Annealing
Elongation
BASE
OH
1
2
3
4
R
Q
F
R
3`
5`

http://dx.doi.org/10.5772/61833

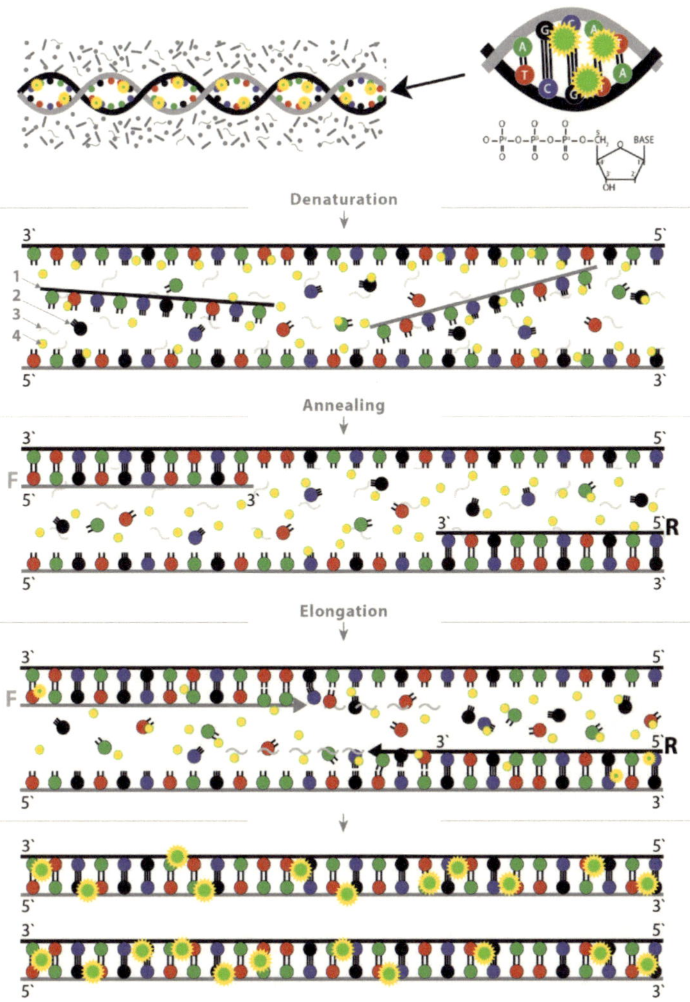

Figure 2. Polymerase chains reactions (PCRs). Components of reactions: 1 – Primers forward and reverse; 2 – deoxynucleoside triphosphates - dATPs, dGTPs dCTP, and dTTPs; 3 – Taq polymerase; 4 – probe labeled with reporter (R) and quencher (Q) molecules; 5 – SYBR Green dye; (A) Conventional PCR includes the following steps: an elongation cycle at 95°C for 1–5 min to activate the polymerase → DNA or cDNA denatures (melts) at 95°C → complementary sequences with specific melting temperature anneal to each other at 40–65°C for 30–120 s → nucleotides are incorporated in the growing strand and the target sequence is amplified at 72°C for up to 60 s. After 35 cycles, a single dsDNA copy is amplified into 2^{36} copies; (B) TaqMan real-time PCR (real-time PCR Taq has 5′–3′ exonuclease activity) 3(b) underlying principle of TaqMan real-time PCR. The reaction mixture includes the same main components as conventional PCR plus a synthetic oligonucleotide (probe) that is labeled with a reporter and a quencher and is complementary to an internal region in the 3′–5′ strand of the sequence of interest (4). The reaction can also include two steps per cycle – denaturation at 95°C and annealing/elongation at 60°C [8, 45]; and (C) Underlying principle of SYBR Green real-time PCR. SYBR Green dye (5) non-specifically binds to dsDNA. During denaturation, the dye is released; it then binds again to the PCR fragment during the elongation step. SYBR Green emits fluorescence when bound to dsDNA. Thus, the more fragments are amplified, the stronger the fluorescence intensity will be. The signal is graphically recorded the same way as in the TaqMan reaction.

Particular attention should be paid to some ambiguities that may arise from the usage of similar acronyms to denote different PCR techniques: the acronym RT is only used to denote reverse transcription; real-time PCR is not abbreviated, and quantitative real-time PCR is commonly denoted as real-time qPCR, whereas real-rime reverse-transcription PCR is typically denoted as real-time RT-PCR, and in quantitative analysis, as real-time RT–qPCR.

Fragment amplification in real-time PCR is based on the same principle as conventional PCR and includes the same basic steps. The difference lies in the method of detection, which needs specially designed equipment. Real-time PCR is based on detection of the fluorescence emitted by a reporter molecule in real time, which is associated with another synthetic oligonucleotide (probe) that is complementary to an internal sequence of the target gene and is labeled with a reporter (R) and a quencher (Q) molecule. The signal emitted by the R molecule is detected after the probe becomes detached from the complementary strand and the R molecule is released by hydrolysis (Figure 2B) – TaqMan version [42]. A signal is emitted and detected in the so-called LightCycler version – by increase and detection of fluorescence resonance energy transfer, via hybridization of R and Q side by side [43–44]. These detection approaches laid the foundations for development of the so-called quantitative real-time PCR (qPCR), which is widely used in infectious disease diagnostics (e.g., human hepatitis viruses), SNP genotyping and allelic discrimination, somatic mutation analysis, copy number detection/variation analysis, chromatin IP quantification, DNA methylation detection, RNA analyses – gene and miRNA expression studies. In this case, the signal is monitored in the course of amplification (i.e., during the early and exponential accumulation of the PCR product) to detect the first significant peak in the amount of PCR product, which is proportional to the initial quantity of target template.

Real-time PCR results are visualized as curves on a graph that reflects the accumulation of signal (Figure 3). The result is obtained based on a pre-prepared standard curve and an internal, positive and negative control that need to be run; i.e., gel electrophoresis is not needed, but may be used as an exception, in case of equipment malfunction, to detect the products.

SYBR Green real-time PCR – This fluorescent dye was first used for detection of nucleic acids in agarose gels [46] and was later introduced in real-time PCR amplifications [44, 46, 47]. The method is based on the fact that the dye **only** binds to double-stranded DNA, which is accompanied with an increase in fluorescence. Thus, the signal intensity correlates with the amount of amplified DNA fragment and, respectively, with the initial sample input amounts (Figure 2C). In 2004, Hubert et al. [48] described the SYBR Green molecule as [2-[N-(3-dimethylaminopropyl)-N-propylamino]-4-[2,3-dihydro-3-methyl-(benzo-1,3-thiazol-2-yl)-methylidene]-1-phenyl-quinolinium]. SYBR Green-based analysis can be used in amplification of any dsDNA and does not require a probe, which makes it less costly. However, the SYBR Green dye may yield false positive signals, as it intercalates into any dsDNA, including non-specific dsDNA sequences.

In general, real-time PCR is more sensitive than conventional PCR and needs the target sequences to be shorter than those used in conventional PCR, maximum 300–400 bp in length; results are obtained in real time and it is not necessary to use gel electrophoresis. The cost of a single reaction (excluding the controls) is much higher than that of conventional PCR. For

http://dx.doi.org/10.5772/61833

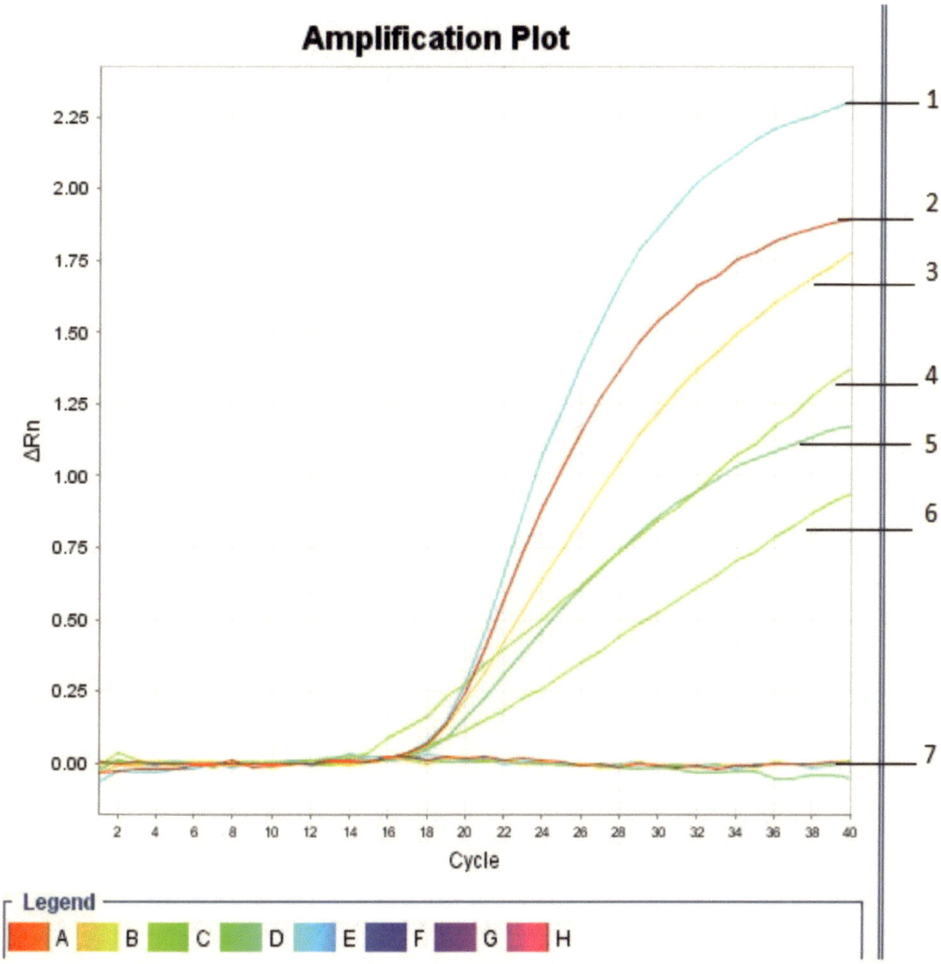

Figure 3. TaqMan real-time PCR data for detection of Shiga toxin genes in DNA extracted from *E. coli* culture broth [45]. 1–6 – amplification curves of internal controls; 7 – lines indicating non-amplification of negative controls and two samples.

example, in conventional PCR, the minimum reaction cost is 0.70 euro, and in TaqMan real-time PCR, about 3 euros (including tips, tubes, and gloves).

Reverse-transcription (RT) PCR is specific in that it includes an additional reverse transcription reaction generating cDNA. This cDNA is then used in conventional or real-time PCR, either in one step (the reaction directly proceeds from reverse transcription to subsequent amplification steps in the same tube) or in two steps (the RT reaction is run separately and a new reaction mixture is prepared for conventional or real-time PCR).

These methods are also applied for multiplex reactions, i.e., amplification of different target sequences in one and the same reaction. What is important for the primer pairs used in the reaction is for them not to be complementary to each other so that they do not form dimers. Another key point is for the primer pairs to have similar annealing temperature (Conventional PCR can tolerate 1°C difference in the annealing temperature of each primer.).

Other methods that are also based on amplification of a target nucleic acid sequence are ligase chain reaction (LCR), nucleic acid sequence-based amplification (NASBA), and strand displacement amplification (SDA).

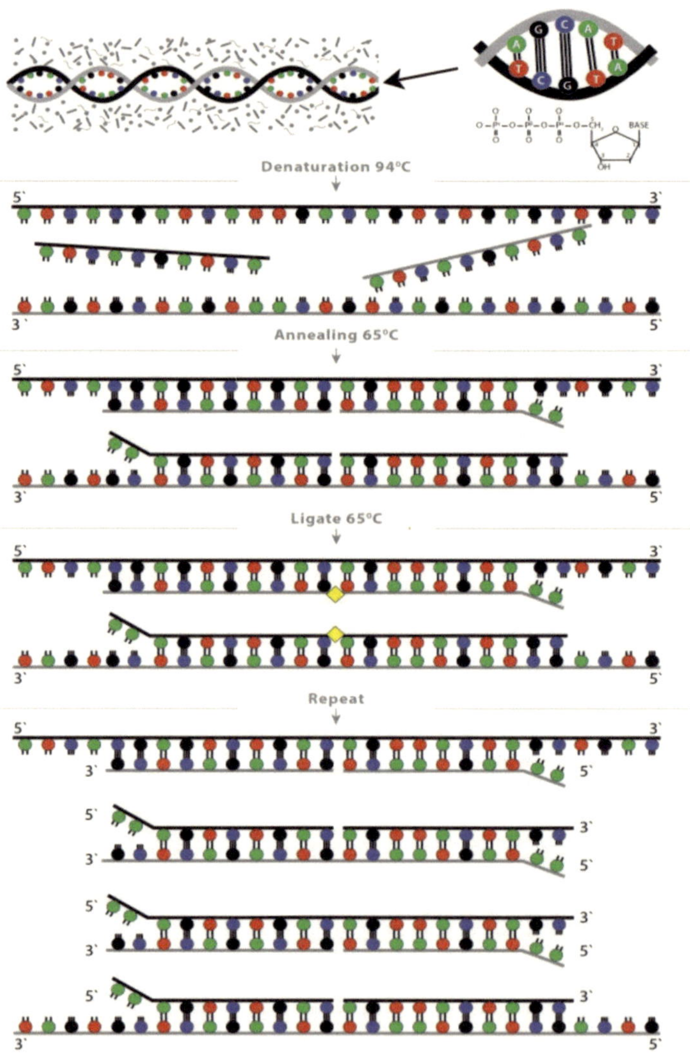

Figure 4. Ligase chain reaction (LCR).

Ligase chain reaction (LCR) was first described by Barany [49]. It combines a ligase reaction with amplification and is particularly suitable for differentiation of single-base substitutions (Figure 4). LCR is based on the following principle: DNA is denatured at 94–95°C and four primers are annealed to the complementary strands at ~65°C, i.e., ca. 5°C below their *Tm*. The thermostable ligase, then, proceeds to ligate only those primers that share perfect complementarity to the target sequence and hybridize immediately next to each other. Thus, if two primers bear a single base-pair mismatch at the junction, they will not ligate effectively enough and, in turn, there will be no product amplification. To avoid ligation of the 3' ends, the discriminating primers contain a 2-bp non-complementary AA tail at their 5' ends [49, 50].

The nucleic acid sequence-based amplification (NASBA) method is based on isothermal amplification used for RNA detection. Briefly, its principle and reaction mixture are as follows: Primer 1, which is complementary to the 3′ end of the target RNA (+) strand and includes a T7 promoter sequence, anneals to it. Next, reverse transcription yields a cDNA (–) strand. Then, the hybrid RNA–cDNA strand is separated via destruction of the RNA strand by RNase H. In the next step, primer 2 anneals to the 3′ end of the DNA (–) strand and reverse transcription yields dsDNA that contains a T7 promoter. Then, T7 RNA polymerase generates RNA (–) copies, primer 2 anneals to the 3′ end of the RNA (–) strand, RT yields copyDNA (+) and primer 1 anneals to copyDNA (+), followed by RT and dsDNA synthesis, which can serve as a template for RNA (+) or RNA (–) synthesis to close the cycle [51].

Strand displacement amplification (SDA) combines the principles of isothermal DNA amplification with those of restriction enzyme digestion [52]. The reaction contains the following components: four primers, DNA polymerase, REase HincII, dGTP, dCTP, dTTP, dATPαS, and takes about 2 h. Basically, in SDA, the primer has two parts: the 5′ end includes a specific HincII enzyme cleavage site (-G-T-T-G-A-C-) and the 3′ end is complementary to the target DNA sequence. DNA polymerase generates a complementary strand and a thiophosphate modification is incorporated – deoxyadenosine 5′-[α-thio]triphosphate (dATP[αS]), in the specific HincII enzyme cleavage site (-C-A_s-A_s-C-T-G-). Then, HincII cleaves the strand at its specific site (-G-T-T ↓ G-A-C-), but not at the complementary sequence (-C-A_s-A_s-C-T-G-). The free 3′-OH group serves as a starting point for strand elongation by DNA polymerase. Thus, the specific HincII cleavage site is regenerated, which closes the cycle.

There are various modifications and versions of these methods, even some commercial kits; however, it is not possible for all of them to be discussed here.

5. Nucleic acid identification by sequencing methods

Sequencing is the basic method to determine the nucleotide sequence of DNA and RNA molecules. There are different sequencing methods and variations, as the methodology evolves based on successful adaptation and application of the properties of nucleic acids.

Sanger method – in 1974, Sanger et al. [53] reported a sequencing method now known as the Sanger method. It is similar with fragment analysis in that the reaction includes a DNA template, radioactive labeled dNTPs and dideoxynucleotides (ddNTPs – without 3′-hydroxyl group, which is essential in phosphodiester bond formation – ddA, ddG, ddC, and ddT), T7 DNA polymerase (with ability to incorporate 2′,3′-dideoxynucleotides), a primer (forward or reverse) and reaction buffer. The annealing, labeling, and termination steps are performed on different thermoblocks, and the polymerase reaction at 37°C. The polymerase enzyme can incorporate either dNTPs or ddNTPs (depending on their relative concentration) at each elongation step. Elongation will proceed if dNTPs are added and will stop if a ddNTP is added at the 3′ end of the strand. The resulting fragments are of different size (length) and, in gel electrophoresis, will migrate toward the positive electrode at a rate of migration inversely proportional to their molecular weight. The method can differentiate fragments that are only 1 bp different in length.

This served as the basis for development of another approach, in which fluorescent dyes are used instead of radioactive isotopes – cycle sequencing or capillary sequencing (the term "capillary" stems from the fact that electrophoresis is performed in a special matrix in capillary tubes, and fluorescence is detected by means of a laser beam). The reaction components are the same as those in the Sanger method, but are mixed in such a way as to allow thermal cycling: denaturation, annealing, elongation, and generation of a balanced population of short and long fragments, using the same principle as the Sanger method (Figure 5A). Specialized software is used to process the detected fluorescence of each fragment and to plot the result as an electropherogram (Figure 5B) in ABI, FASTA, and PAUP file format.

A limitation of the method is the size of the fragments that can be sequenced: maximum 800–1000 nucleotides per run.

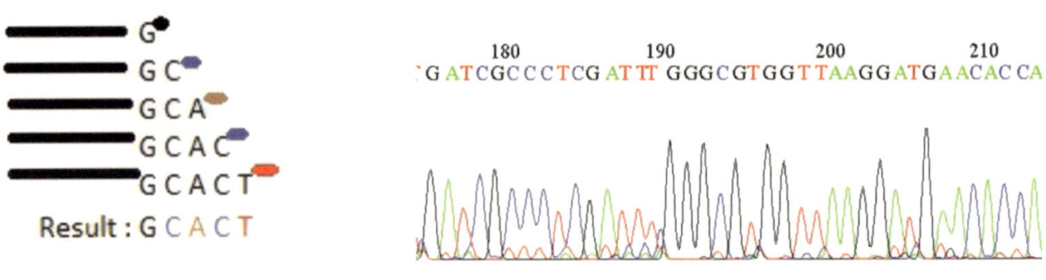

A B

Figure 5. Generation of fragments by incorporation of dNTPs and ddNTPs (labeled) in the Sanger method and cycling sequencing (A). Sample electropherogram (portion) of the beta-lactamase OXA-48 gene of *Klebsiella pneumoniae* strain OXA48BG, NCBI, GenBank accession number KJ959619.1 [54] (B).

The sequences obtained using forward and reverse primers are analyzed by programs available either as freeware [55, 56] or as commercial software. What is important to remember is to always check the sequencing results in the file generated by the software against the electropherogram: there may often be discrepancies between the nucleotide sequence in the file and that in the electropherogram. In such cases, the electropherogram should be considered more reliable but the background effect should also be accounted for.

In the alignment of sequences (first, between the F and R primers of a sample and, second, between different samples), it is important for them to be equal in length. The next step is sequence analysis – phylogenetic analysis, genetic distance analysis, etc. For example, in phylogenetic analysis, it is particularly important to choose the mathematical model that is most appropriate for each particular case. Instead, there are software programs especially designed to determine the most appropriate model depending on the sequence length, the number of sequences, potential substitutions, etc. For example, jModelTest [57] analyzes and selects among 89 different models.

5.1. Next-Generation Sequencing (NGS) platforms

The way is now open for nucleic acid research using NGS technologies. This part of the chapter will focus on three NGS platforms: Illumina, Nanopore, and Ion Torrent.

5.2. Illumina sequencing by synthesis

The Illumina technology is based on the principle described above, i.e., incorporation of fluorescently labeled dNTPs by DNA polymerase using a DNA or a cDNA template in a sequence of cycles. The identification of incorporated nucleotides is based on fluorophore excitation. The whole process includes the following steps:

- Nucleic acid preparation is one of the critical steps (it may vary depending on the operator and the consumables/method used). It is important for the target DNA (total DNA, PCR products, cDNA) to be of the highest possible purity.
- Sequencing library preparation (another critical step) is done by random fragmentation of the target DNA, followed by 5′ and 3′ ligation of fragments by adapters. Thus, each strand in the DNA double helix is adapter-ligated. Fragmentation and ligation may be performed in a single step, which, according to (9), markedly increases the effectiveness. The adapter-ligated fragments are then amplified by PCR and purified in a gel.
- Cluster generation takes place on the surface of a flow cell. The flow cell is, in fact, a glass slide with channels on whose surface two types of oligonucleotides are attached. These oligonucleotides are complementary to one of the ends of the library fragments (the adapter) and are identical with the other end. This ensures cluster generation is done by hybridization of the adapter-ligated amplified fragments from the library with synthetic oligonucleotides on the surface of the flow cell that are complementary to the adapters. Each single-stranded fragment binds with the complementary end, while its other end remains free, perpendicular to the surface of the flow cell (The free end is identical with the oligonucleotide in the flow cell to provide synthesis of a complementary sequence in the new strand.). Then complementary strand synthesis can begin, displacing the original strand when the synthesis is completed. The fragments thus bound act as templates for the so-called bridge amplification of discrete clonal clusters. In bridge amplification, the free end of a new single-

stranded DNA fragment anneals (by hybridization) to a complementary oligonucleotide on the flow-cell surface (resulting in an ∩-shaped bridge-like structure); then, a DNA polymerase enzyme synthesizes a complementary DNA strand by incorporation of unlabeled nucleotides. The resulting double-stranded DNA template is denatured, which releases the two single-stranded DNA fragments both from one another as well as from the flow-cell surface at one end: if the initial ssDNA fragment had its 5′ end free, then it is restored and the new copy remains bound at its 3′ end (which is complementary to the free 5′ end). Thus, the templates generated by bridge amplification form clusters of single-stranded fragments bound at one end and the number of DNA templates grows. (Then, the reverse templates are removed and only the forward strands remain – clones identical to the original templates)

- Sequencing – The samples are ready for the sequencing step when the clusters are complete. A proprietary reversible terminator-based method is used: a sequencing primer is annealed (it is complementary to the part of the adapter that is adjacent to the DNA fragment of interest), followed by cycles of incorporation of terminator-bound dNTPs. In each cycle, there is a mixture of the four different dNTPs and they naturally compete for incorporation based on complementarity. As a result, only one complementary dNTP per cycle is incorporated in the growing strand. In each cycle, a labeled terminator-bound dNTP is incorporated in each chain in each cluster and is detected based on emission wavelength and intensity. The signal emitted by each nucleotide incorporated in each strand at the end of a cycle is recorded in real time in each cluster. The cycle is repeated n times to obtain an n-base long sequence. When the new strand is synthesized and its sequence is recorded, it is removed. This approach of base-by-base sequencing achieves very high precision, as errors are eliminated even in repetitive sequence regions and homopolymers.

- Data processing and analysis – The fluorescence emissions from different clusters are digitally processed in parallel and are visualized as nucleotide sequences for each individual cluster. Following sequence alignment, the data are compared against referent sequences to perform phylogenetic analysis, single-nucleotide polymorphism analysis, distance determination, and a range of other analyses.

This technique allows multiplex analysis to be carried out by creating distinct libraries based on the so-called index sequences (short nucleotide sequences specific for each library that can be used like a barcode), which are attached during library preparation step. Different libraries are first individually prepared and are then combined together and loaded in one and the same flow cell lane. The labeled libraries are sequenced simultaneously, in a single run of the equipment, and at the end of the process, the sequences are exported in a single file. Next, a demultiplexing algorithm is used to separate the sequences in different files based on their barcode. This is followed by alignment with referent sequences of interest.

The platform is compatible with different library preparation methods depending on the purpose of the sequencing analysis (whole genome sequencing, target sequencing, mRNA sequencing, 16s RNA gene sequencing, etc.). This is possible because the sequencing steps that come after the library preparation step are fundamental and do not depend on the library preparation method. There are ready-to-use, standardized library preparation kits designed for different sequencing purposes and more and more new approaches and modifications are being developed [58].

5.3. Nanopore technology

Nanopore technology is an innovative NGS platform that also includes sequence library preparation. The library preparation step requires maximal purity of the target nucleic acid (DNA, PCR products, cDNA). This is achieved by magnetic particle purification (It is probably possible to use other methods as well, e.g., column silica-based ones, if the manufacturers are reliable and guarantee high purification of the nucleic acid samples). The target nucleic acid samples are ligated between the 5′ end of one of the strands and the 3′ end of the other strand of the double-stranded DNA (in the shape of a hairpin). Bayley sequencing technology is used. In this method, the DNA molecule is basically passed through a system consisting of a processive exonuclease enzyme bound to a protein nanopore (Figure 6A). The enzyme unzips the DNA double helix and pushes/cleaves one of the strands through the aperture of the protein nanopore in a base-by-base manner (The aperture is only a few nanometers in diameter.). This continues until the hairpin loop at the end of the first strand is reached and then the enzyme simply proceeds on to push the ligated reverse strand through the nanopore as well. The system is designed so that a constant ion current flows through the aperture of the free protein nanopore. This current is specifically disrupted when each base enters the aperture. These specific disruptions in the ion current are recorded by an electronic device and are interpreted to identify each base. The disruption that each of the four bases causes in the ion current is different due to differences in their chemical structure and chemical characteristics. Recordings from multiple channels in parallel allows high-throughput sequencing of DNA.

Protein nanopores are inserted into an electrically resistant polymer membrane so that a membrane potential can be created driving the ion current through the pore aperture. This ion current is disrupted in a specific way when a nucleotide passes through the aperture of the protein nanopore, which causes a change in the membrane potential.

An array of microscaffolds holds the membrane in which the nanopores are embedded, giving stability to the structure during operation. Each microscaffold on the sensor array chip contains an individual electrode, allowing for multiple nanopore experiments to be performed in parallel. Each nanopore channel is controlled and measured by an individual channel on a corresponding, bespoke application specific integrated circuit (ASIC) (Figure 6B) [59].

Another NGS platform is the Ion Torrent sequencing technology. In this method, following library preparation, the sequencing step is actually performed in wells that contain the DNA template, an underlying sensor, and electronics. It works as follows: when a new nucleotide is incorporated in the growing DNA strand, a proton (H^+) is released, which causes a change in the pH in the well. This leads to changes in the surface potential of a metal-oxide sensing layer and to changes in the potential of the source terminal of the underlying field-effect transistor. This is the signal that a complementary nucleotide has been incorporated at the end of the growing DNA strand. To determine the DNA sequence, the equipment needs to be able to differentiate between the four bases. This is done by flushing each well with one type of nucleotide at a time. If the base is complementary, the nucleotide will be incorporated and a

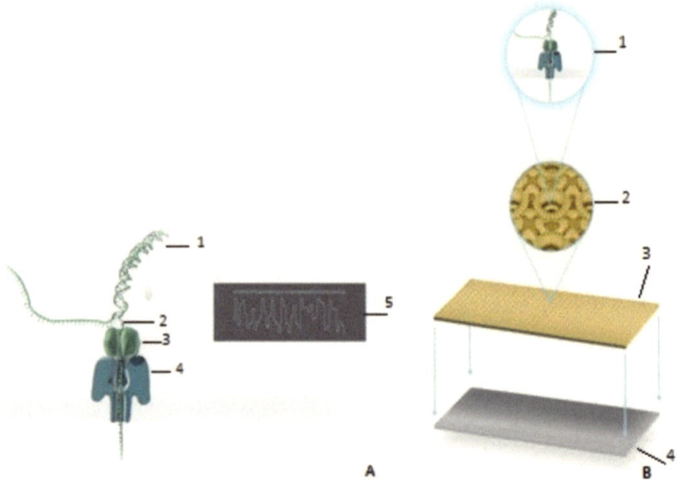

Figure 6. (A) Underlying principle used in nanopore technology sequencing (illustrative data). 1 – DNA library is prepared; 2 – the dsDNA strands are separated and one of the strands passes through 3 – a processive exonuclease enzyme, and 4 – a protein nanopore, in a base-by-base manner, causing a characteristic disruption in the electrical current flowing through the nanopore, which is in fact the signal that is detected in real time. When the end of the first strand is reached, the 3′- to 5′-end hairpin ligation of the two DNA strands allows the processive enzyme to simply continue to push the second strand through the nanopore; 5 – Representative image of DNA sequencing data. (B) 1 – A nanopore protein; 2 – Array of Microscaffold; 3 – Array Chip; 4 – Application Specific Integrated Circuit.

change in the pH will be recorded; and if the base is not complementary, there will be no pH change [60].

NGS technologies are a new trend and yet more approaches and applications in nucleic acid analysis are being developed.

Author details

Ivo Nikolaev Sirakov*

Address all correspondence to: insirakov@gmail.com

Medical University – Sofia, Department of Microbiology, Sofia, Bulgaria

References

[1] Fabre AL, Colotte M, Luis A, Tuffet S, Bonnet J. An efficient method for long-term room temperature storage of RNA. Euro J Human Genetics 2014;22:379–85; doi: 10.1038/ejhg.2013.145

[2] Poulsen HE, Specht E, Broedbaek K, Henriksen T, Ellervik C, Mandrup-Poulsen T, Tonnesen M, Nielsen PE, Andersen HU, Weimann A. RNA modifications by oxidation: a novel disease mechanism? Free Radic Biol Med 2012;52:1353–61.

[3] Bruskov VI, Malakhova LV, Masalimov ZK, Chernikov AV. Heat-induced formation of reactive oxygen species and 8-oxoguanine, a biomarker of damage to DNA. Nucleic Acids Res 2002;30:1354–63.

[4] Evans RK, Xu Z, Bohannon KE, Wang B, Bruner MW, Volkin DB. Evaluation of degradation pathways for plasmid DNA in pharmaceutical formulations via accelerated stability studies. J Pharm Sci 2000;89:76–87.

[5] Cataldo F. Ozone degradation of ribonucleic acid (RNA). Polym Degrad Stabil 2005;89:274–81.

[6] Seyhan AA, Burke JM. Mg^{2+}-independent hairpin ribozyme catalysis in hydrated RNA films. RNA 2000;6:189–98.

[7] Butler E, Gelbart T, Kuhl W. Interference of heparin with the polymerase chain reaction. BioTechniques 1990;9:166.

[8] Identification of the subtypes of Verocytotoxin encoding genes (vtx) of *Escherichia coli* by conventional PCR - EU Reference Laboratory for E. Coli, Istituto Superiore di Sanità, EU-RL VTEC_Method_006_Rev 1; 18.06.2013.

[9] Sirakov I, Popova R, Daskalov H, Slavcheva I, Gyurova E, Mitov B. Shigatoxin-producing Escherichia coli in raw cow milk from small farm producers and phylogenetic subtype determination. JFAE 2014;12(3,4):108–14.

[10] Popova R [PhD thesis] Research on spread, characteristics and pathogenic factors of Escherichia coli isolated from food. Sofia, Bulgaria: National Diagnostic and Research Veterinary Institute, Bulgarian Food Safety Agency; 2014.

[11] Volkin E, Carter CE. The preparation and properties of mammalian ribonucleic acids. J Am Chem Soc 1951;73(4):1516–9.

[12] Chirgwin JM, Przybyla AE, MacDonald RJ, Rutter WJ. Isolation of biologically active ribonucleic acid from sources enriched in ribonuclease. Biochemistry 1979;18(24): 5294–9.

[13] Chomczynski P, Sacchi N. Single-step method of RNA isolation by acid guanidinium thiocyanate-phenol-chloroform extraction. Anal Biochem 1987;162:156–9.

[14] Perry RP, La Torre J, Kelley DE, Greenberg JR. On the lability of poly(A) sequences during extraction of messenger RNA from polyribosomes. Biochim Biophys Acta 1972;262(2):220–6.

[15] Palmiter RD. Magnesium precipitation of ribonucleoprotein complexes. Expedient techniques for the isolation of undegraded polysomes and messenger ribonucleic acid. Biochemistry 1974;13:3606–15.

[16] Brawerman G, Mendecki J, Lee SY. A procedure for the isolation of mammalian messenger ribonucleic acid. Biochemistry 1972;11(4):1972.

[17] Puissant C, Houdebine LM. An improvement of the single-step method of RNA isolation by acid guanidinium thiocyanate – phenol – chloroform extraction. BioTechniques 1990;8:148–9.

[18] Birnboim HC1. Extraction of high molecular weight RNA and DNA from cultured mammalian cells. Methods Enzymol 1992;216:154–60.

[19] Rio DC, Ares M Jr, Hannon GJ, Nilsen TW. Purification of RNA using TRIzol (TRI reagent). Cold Spring Harb Protoc 2010 Jun;2010(6):pdb.prot5439. doi: 10.1101/pdb.prot5439

[20] Wyatt GR. The purine and pyrimidine composition of deoxypentose nucleic acids. Biochem J 1951;48(5):584–90.

[21] Clark LB, Tinoco I Jr. Correlations in the ultraviolet spectra of the purine and pyrimidine bases. J Am Chem Soc 1965;87(1):11–5.doi: 10.1021/ja01079a003

[22] Barbara R, Karin F, Claus V, Martin W, Peter R. Impact of long-term storage on stability of standard DNA for nucleic acid-based methods. J Clin Microbiol Nov 2010;4260–2. doi:10.1128/JCM.01230-10

[23] Bonner G, Klibanov AM. Structural stability of DNA in nonaqueous solvents. Biotechnol Bioeng 2000;68:339–44.

[24] Haralambiev H. Animal Viruses. Diagnostic of virus infection. Publisher – Pandora, Haskovo, 2002, pp. 58–59.

[25] Watson JD, Crick FHC. A structure for deoxyribose nucleic acid. Nature 1953;171:737–8.

[26] Langer PR, Waldrop AA, Ward DC. Primer-directed enzymatic amplification of DNA with a thermostable DNA polymerase. Science 1988;239:487. doi: 10.1126/science.2448875

[27] Saiki RK,Gelfand DH,Stoffel S, Scharf SJ, Higuchi R,Horn, GT,Mullis KB, Erlich H. Primer-directed enzymatic amplification of DNA with a thermostable DNA polymerase. Science 1988;239(4839):487–91.

[28] Linz U, Delling U, Rubsamem W. Systematic studies on parameters influencing the performance of the polymerase chain reaction. J Clin Chem Clin Biochem 1990;28:5–13.

[29] Sarkar G, Kapelner S, Sommer SS. Formamide can dramatically improve the specificity of PCR. Nucleic Acids Res 1990;18:7465–8.

[30] He Q, Marjamäki M, Soini H, Mertsola J, Viljanen MK. Primers are decisive for sensitivity of PCR. Biotechniques 1994;17(1):82, 84, 86–7.

[31] Drummond R, Gelfand DH. Isolation, characterization and expression in *Escherichia coli* of the DNA polymerase gene from *Thermus aquaticus*. J Biol Chem 1989;264:6427–36.

[32] Vinay KS, Anil K. PCR primer design. Molecul Biol Today 2001;2(2):27–32.

[33] Sheffield VC, Cox DR, Lerman LS, Myers RM. Attachment of a 40 base pair G+C rich sequence (GC-clamp) to genomic DNA fragments by polymerase chain reaction results in improved detection of single base changes. Proc Natl Acad Sci 1989;86:232–6.

[34] Taylor JM, Ilmensee IL, Summers J. Efficient transcription of RNA into DNA by avian sarcoma virus polymerase. Biochim Biophys Acta 1976;442:324–30.

[35] Noonan KE, Roninson IB. mRNA phenotyping by enzymatic amplification of randomly primed cDNA. Nucl Acids Res 1988;16(21):10366.

[36] Feinberg AP, Vogelstein B. A technique for radiolabeling DNA restriction endonuclease fragments to high specific activity. Anal Biochem 1983;132(1):6–13.

[37] Jones C, Kortenkamp A. RAPD library fingerprinting of bacterial and human DNA: applications in mutation detection. Teratog Carcinog Mutagen 2000;20(2):49–63.

[38] Rossetti L, Giraffa G. Rapid identification of dairy lactic acid bacteria by M13-generated, RAPD-PCR fingerprint databases. J Microbiol Methods 2005;63(2):135–44.

[39] Kumar NS, Gurusubramanian G. Random amplified polymorphic DNA (RAPD) markers and its applications. Sci Vis 2014;11(3):116–24.

[40] Cowan JA. Structural and catalytic chemistry of magnesium-dependent enzymes. Biometals 2002;15(3):225–35.

[41] Sreedhara A, Cowan JA. Structural and catalytic roles for divalent magnesium in nucleic acid biochemistry. Biometals 2002;September 15,3:211–23.

[42] Livak KJ, Flood SJA, Marmaro J. Method for detecting nucleic acid amplification using self-quenching fluorescence probe. Applied Biosystems Division, Perkin-Elmer Corp. US Patent US5538848, 1996.

[43] Cardullo A, Agrawal S, Flores C, Zamecnik PC, Wolf DE. Biochemistry detection of nucleic acid hybridization by nonradiative fluorescence resonance energy transfer. Proc Natl Acad Sci USA 1988; Dec 85:8790–4.

[44] 44 multisample fluorimeter with rapid temperature control. Biotechniques 1997;22:176–81.

[45] Popova R, Taseva R, Sirakov I, Daskalov H. Fecal contamination of natural water resevoirs in Pirin Mountain with Escherichia coli. Tradition and Modernity in Veterinary Medicine. Issn 1313-4337, University Of Forestry, Sofia, Bulgaria, pp. 98–102, 12/2012.

[46] Schneeberger C, Speiser P, Kury F, Zeillinger R. Quantitative detection of reverse transcriptase–PCR products by means of a novel and sensitive DNA stain. PCR Methods Appl 1995;4:234–8.

[47] Bengtsson M, Karlsson HJ, Westman G, Kubista M. A new minor groove binding asymmetric cyanine reporter dye for real-time PCR. Nucleic Acids Res 2003;31:45.

[48] Giglio S, Monis PT, Saint CP. Demonstration of preferential binding of SYBR Green I to specific DNA fragments in real-time multiplex PCR. Nucleic Acids Res. 2003;31:136.

[49] Zipper H, Brunner H, Bernhagen J, Vitzthum F. Investigations on DNA intercalation and surface binding by SYBR Green I, its structure determination and methodological implications, Nucl Acids Res 2004;32:12, 103. doi:10.1093/nar/gnh101

[50] Barany F. The ligase chain reaction in a PCR world. Genome Res 1991;1:5–16.

[51] Wiedmann M, Wilson WI, Czajka J, Luo J, Barany F, Batt CA. Ligase chain reaction (LCR)--overview and applications. Genome Res 1994;3:S51–S64.

[52] Guatelli JC, Whitfield KM, Kwoh DY, Barringer KJ, Richman DD, Gingeras TR. Isothermal, in vitro amplification of nucleic acids by a multienzyme reaction modeled after retroviral replication. Proc Natl Acad Sci USA 1990;87:1874–8.

[53] Walker GT, Little MC, Nadeau JG, Shank DD. Isothermal in vitro amplification of DNA by a restriction enzyme/DNA polymerase system. Proc Natl Acad Sci USA 1992;89:392–6.

[54] Sanger F, Donelson JE, Coulson AR, Kösse H, Fischer D. Determination of a nucleotide sequence in bacteriophage f1 DNA by primed synthesis with DNA polymerase. J Mol Biol 1974;90(2):315–33.

[55] Marteva-Proevska Y, Strateva T, Sirakov I, Zlatkov B, Markova B. Emergence of *Klebsiella pneumoniae* carrying OXA-48 carbapenemase in Bulgaria. International Journal of Biological Sciences and Applications 2015 x(x): xx-xx

[56] Thompson JD, Higgins DG, Gilbson TJ. CLUSTAL W: improving the sensitivity of progressive multiple sequence alignment through sequence weighting, position-specific gap penalties and weight matrix choice. Nucleic Acids Res 1994;22:4673–80.

[57] Edgar RC. Muscle: multiple sequence alignment with high accuracy and multiple sequence alignment with high accuracy. Nucleic Acids Res 2004;32(5):1792–7.

[58] Posada D. J Model Test: phylogenetic model averaging. Mol Biol Evol 2008;25:1253–6.

[59] Illumina Nextera DNA Library Preparation Kits data sheet. 2004. www.illumina.com/documents/products/datasheets/datasheet_nextera_xt_dna_sample_prep.pdf

[60] https://nanoporetech.com/applications/dna-nanopore-sequencing

[61] https://www.thermofisher.com/us/en/home/life-science/sequencing/next-generation-sequencing/ion-torrent-next-generation-sequencing-technology.html

Chapter 2

Nucleic Acids Extraction from Formalin-Fixed and Paraffin-Embedded Tissues

Gisele R. Gouveia, Suzete C. Ferreira, Sheila A. C. Siqueira and Juliana Pereira

Additional information is available at the end of the chapter

http://dx.doi.org/10.5772/61581

Abstract

Formalin-fixed paraffin-embedded (FFPE) tissues are an important sample source for retrospective studies. Despite its ability to preserve proteins and cell morphology, formalin hinders molecular biology tests since it fragments and chemically modifies nucleic acids, especially RNA. Although several studies describe techniques that allow extracting nucleic acids from FFPE tissues, so far there is no consensus in the literature about the best protocol to be used in this type of material. Thus, the current chapter aims to describe the factors affecting the FFPE tissue nucleic acid extracting process, compare the available protocols and to describe the modifications developed by our group in some protocols. Such modifications enable nucleic acids obtainment in satisfactory quantity and quality for molecular biology studies.

Keywords: DNA, RNA, FFPE, extraction

1. Introduction

Formalin-fixed paraffin-embedded (FFPE) tissues are of great importance to retrospective studies. Their main advantage lies on the possibility of correlating genetic and molecular biology analyses with data from patients' medical records and clinical outcomes [1].

Formalin fixation is the most widely used method for tissue fragment preservation. It is a low-cost and easy-to-handle method, which preserves good morphological cell quality. The method is compatible with the antibodies used in the immunohistochemistry technique [2]. However, although formalin fixation and routine histological processing techniques preserve tissue cellular morphology and protein integrity, they also impair the obtainment of nucleic acids with the same quality, especially RNA, since they degrade and chemically modify such acids [3, 4].

Formalin replacement by other tissue fixatives such as Bouin, Carnoy, alcohol or HOPE (glutamic acid buffer Hepes-mediated organic solvent effect protection) may be an alternative to reverse the problem [4, 5]. However, since formalin is the fixative of choice in most pathology departments, several research groups have sought to reverse the chemical changes caused by this fixative type.

Although several papers describe techniques that allow FFPE tissue nucleic acid extracting process [6–8, 3, 1, 4], so far there is no consensus in the literature about the best protocol to be used in this type of material.

The studies based on such approach not often detail the used methodology. Besides, not all the published techniques were reproduced by our group. It took us approximately 8 months to standardize the DNA and RNA extraction process in our FFPE tissues research.

The current chapter aims to describe the factors affecting the process of nucleic acid extraction from FFPE tissue, compare the available protocols and to describe the modifications developed by our group in some protocols that enable nucleic acids obtainment in satisfactory quantity and quality for molecular biology studies.

2. Factors affecting nucleic acids extraction from FFPE tissues

Difficulties in obtaining quality nucleic acids, especially RNA, cause degradation and chemical modification, despite the fixation in formalin and the routine histological processing techniques used to preserve cellular morphology and tissues' protein integrity [3–4].

Nucleic acids extraction from FFPE tissue shows some critical issues that may affect the quality of the obtained DNA or RNA as well as these samples' subsequent amplification process. Such critical matters include tissue fixing and clamping as well as the post-fixing stage (tissue cutting preparations, deparaffinization and hydration processes, in addition to other stages of the extraction process itself, such as digestion and purification), which will be described below [9].

Pre-fixation is defined as the period between tissue collection and the beginning of the setting process. The material starts degrading shortly after its collection, right when the tissue is exposed to hypoxia and to the DNases and RNases found in the environment. The first biochemical modifications emerge after 10 min of anoxia. Thus, it is very important to reduce the pre-clamping time to seconds [4, 9].

The fixing conditions (time, temperature and fixative type) and, in some cases, the descaling processes alter material preservation and directly influence the quantity and quality of the obtained nucleic acid. Fixations kept in formalin solution for more than a week destroy the nucleic acids and lead to the cross-linking of all tissue components. It results in highly fragmented nucleic acids, which are more resistant to the extraction process [2, 3, 9].

http://dx.doi.org/10.5772/61581

Chemical studies have shown that formaldehyde breaks hydrogen bonds in double-stranded DNA adenine- and thymine-rich regions. It creates new chemical interactions in protein folding, thus resulting in bonds between DNA proteins and DNA fragmentation [10].

RNA messenger (mRNA) obtained from FFPE tissues is often not intact. It is usually degraded to less than 300 base pairs [1]. However, Hamatani et al. (2006) [11] found that 80% of the RNA samples presenting 60 base pairs may be sufficiently amplified by polymerase chain reactions (PCR). All post-fixation stages, such as the paraffin blocks attainment sections, are also essential to the obtainment of high-quality nucleic acids.

Contamination is one of the critical issues affecting the quality of the samples. Thus, it is necessary to decontaminate the workstation as well as to use DNases- and RNases-free tools. These DNases and RNases result from paraffin block cuts used to extract the nucleic acid of interest [2].

Although some authors state that deparaffinization is not a required step [2], most protocols suggest that the material must be deparaffinized before the extraction process in order to obtain nucleic acids in a more efficient way. Most protocols use solvent (usually xylene) to remove the paraffin cuts, and this procedure is followed by ethanol-based rehydration.

No matter the used protocol, digestion is the first step in the nucleic acid extraction process, and it aims to lyse membranes in order to release the cellular components. This step may be accomplished by several methods, such as elevated temperatures, enzymatic digestion, mechanical disruption or even by using other detergents or according to cell type solutions. In general, enzymatic digestion with proteinase K is used in most protocols; however, the concentrations and the incubation times are highly variable [9].

Nucleic acids purification is the next stage. Literature reports protocols using organic solvents (such as phenol-chloroform) [12–14], salt (salting out) [15] and other substances (Chelex-100) [16] as well as protocols using commercial kits available in the market [9].

Li et al. (2008) [1] observed that RNA extraction protocols based on proteinase K digestion followed by DNase, column purification and elution treatments led to good results in FFPE samples. These authors have shown that proteinase K is essential to degrade covalently linked proteins in order to release RNA from the cell array and to inactivate RNAses, which tend to be stable.

Ribeiro-Silva and Garcia (2008) [4] have shown that proteinase K is used to degrade proteins bound to nucleic acids and that the incubation between 60°C and 70°C removes the methylol groups added by formalin. RNA isolation by denaturing agents prevents the RNases action. In addition, deoxyribonuclease (DNase) incubation is required to remove the deoxyribonucleic acid (DNA) sample. Finally, purification by precipitation with alcohol porous column removes any residue and contaminants.

All tested protocols will be detailed in the sections below, as well as the changes suggested to determine the protocols that would be viable to the obtainment of nucleic acids presenting adequate quality for molecular biology studies.

3. Preparing FFPE tissue sections for nucleic acids extraction

All samples included in the studies conducted by our team result from biopsies performed in diffuse large B cell lymphoma patients, from lymph node samples or from reaction amygdala samples stored in the Pathology Department of Hospital das Clínicas, Medical School of University of São Paulo. All the herein used samples were formalin-fixed and paraffin-embedded (FFPE) according to the standard methods described in the literature.

Four 20-μm thick cuts were performed in each sample using routine histological techniques. The sections were placed into 1.5 mL RNase- and DNase-free microtubes, and they were subsequently subjected to RNA and DNA extraction processes.

The first protocol used to prepare the cuts in the nucleic acid extraction process (suggested for the majority of commercial kits and protocols) consisted of deparaffinizing the sections with xylene and of rehydrating them with ethanol. In order to do so, 1 mL xylene PA was added to each sample (Synth® Diadema, SP, Brazil), and it was followed by homogenization using Vortex Genie 2T (Scientific Industries, Inc., Bohemia, NY, USA) and by incubation at 50°C, for 5 min, in digital thermomixer (Eppendorf AG, Hamburg, Germany). After incubation, the samples were centrifuged at maximum speed for 5 min in the R-5418 microfuge (Eppendorf AG, Hamburg, Germany). The xylene was discarded and the cell button was washed two times with absolute ethanol (Merck KGaA, Darmstadt, Germany). The supernatant was discarded after each wash. After the cell button was completely dried, the extraction process started, as it is described in the following sections.

Moreover, even after some RNA extraction methods that allowed finding some viable options were compared, it was observed that not all the extracted samples showed successful amplification in PCR reactions [17].

None of the previous studies available in the literature described the possible factors that could influence the amplification success. Therefore, the current study made the option of investigating some of the potential interferences in the nucleic acid obtainment process, namely: tissue fragment size, blocks' storage time, used fixative type, different cDNA synthesis primers and different primer sequences, among others [18].

After the aforementioned study, the tissue preparation protocol for RNA extraction process was modified by including a washing step. It consisted of using 1 ml phosphate buffer saline (PBS) with 5 min incubation at room temperature, followed by full speed centrifugation for 5 min in the R-5418 microfuge (Eppendorf AG, Hamburg, Germany). This procedure is done to remove possible fixative residues that could work as PCR inhibitors [18]. Next, the same process was used in the DNA extraction performed by our team.

Such protocol change was done under the assumption that the amplification failure in PCR reactions could be caused by the presence of contaminants such as fixative waste working as PCR inhibitors.

The results show that the PBS washing step inclusion in the samples' extraction preparation process led to some statistically significant advantages such as the obtainment of better

RNA concentration results (p = 0.00025), even when the same initial amount of tissue was used. In addition, the washing step allowed obtaining better sample purity levels (p = 0.0000001), increasing the samples amplification success (p = 0.018) both in the standard and in the real-time PCR reactions [18].

Two possible factors may have influenced the improved amplification efficiency of these samples. The first is based on the fact that formalin is water-soluble, thus the PBS washing step may have led to tissue-fixation residues' solubilization and removal, and it could possibly work as PCR inhibitors. Furthermore, previous studies suggested that pH values between 6.5 and 9.0 are optimal for amplification [11, 9]. Then, the PBS solution may have possibly altered pH levels, thus increasing the quality of the obtained RNA and the amplification success.

Despite the influence of the PBS-based washing step addition on the samples' successful amplification, we observed the fixative interference in this process too. Formalin-fixed samples showed more successful PCR amplification reactions than those fixed in formaldehyde or in Bouin's solution, even after further PBS washing (p = 0.000018) [18].

Another fact observed by our team refers to the paraffin-embedded tissue fragment size. All fragments size equals to or greater than 1.0 cm had the most successful samples' amplification (p = 0.034) [18]. This may be possibly due to the fact that smaller tissues may increase the fixative absorption and, consequently, cause greater nucleic acids degradation and chemical modification in the full extent of the tissue.

No impact was found on the amplification success when the other variables suggested in our study were tested (tissue type, block age, different primers used in the cDNA synthesis or the endogenous genes used in the PCR reaction), regardless of the use (or not) of the PBS washing step [18].

This protocol change was kept in the preparation of samples subjected to DNA extraction, due to the statistically significant impact caused by the PBS washing step inclusion on either the obtainment of better RNA concentrations or purity relations.

It is possible to conclude that this sample preparation is an essential step to obtain better quality nucleic acids for molecular biology studies.

4. RNA extraction from FFPE tissues

Three different RNA extraction protocols were tested, as described below:

Protocol 1: Commercial kit RECOVERALL Total Nucleic Acid Isolation Optimized for FFPE Samples (Ambion, Inc., Austin, Texas, USA) [17–18]

Two hundred (200μ l) of digestion buffer and 5 μL protease were used for tissue lysis in each sample. It was followed by incubation for 15 min at 50°C and for 15 min at 80°C. Next, RNA isolation was held by the addition of 790 μL buffer containing absolute ethanol and by the subsequent passage through the separation column. After washing the column with two wash buffers, realized treatment with DNAse and further washing with two buffers, according to the manufacturer's instructions. The eluted RNA was obtained by using 60 μL of the kit elution buffer at RT. After incubation for 5 min at room temperature, the samples were centrifuged at maximum speed and the obtained RNA was stored at –80°C until its use.

Protocol 2: Paradise®Whole Transcript RT Reagent kit System (Arcturus Bioscience, Inc., Mountain View, California, USA) [17]

Incubation with buffer containing proteinase K was held for 20 h, at 37°C, after the samples' preparation process, to digest the proteins. The RNA isolation was performed by two successive washes using buffer containing the ethanol kit. Then, the samples were purified by passing them through the column kit according to the manufacturer's instructions. Subsequently, samples were incubated with buffer containing DNase at 37°C for 15 min and at 4°C for 1 min. DNase inactivation was performed by incubating the samples at 70°C for 10 min and at 4°C for 1 min. The RNA samples were stored at –80°C until they were used.

Protocol 3: Trizol extraction method (Invitrogen, UK) [17]

The RNA extraction by Trizol method was performed as recommended by Körbler et al. (2003) [8] and Antica et al. (2010) [19]. The tissue was digested by incubating the samples in buffer containing 10 mM NaCl, 500 mM Tris pH 7.6, 20 mM ethylenediaminetetraacetic acid (EDTA), 1% sodium dodecyl sulfate (SDS) and 500 μg/mL proteinase K. First, the sample was incubated at 55°C, for 3 h; then, it was incubated at 45°C overnight with proteinase K inactivation by elevating the temperature to 100°C for 7 min in the next day. Finally, all samples were subjected to RNA extraction process according to the classic Trizol method previously described by Chomczynski and Sacchi (1988) [20]. The obtained RNA was stored at –80°C until it was used.

4.1. Evaluating RNA concentration and quality

NanoDrop equipment (NanoDrop Technologies, Inc. Wilmington, DE) was used to evaluate the concentration and purity of RNA samples extracted according to the three protocols described above. RNA amounts above 50 ng/μl with purity between 1.7 and 1.9 were considered to be suitable.

4.2. Results and comments on the herein described RNA extraction protocols

The purity levels and degrees obtained by the three protocols were satisfactory (over 50 ng/μL and purity concentrations between 1.8 and 1.9). However, only samples obtained according to protocols 1 (Ambion) and 2 (Arcturus Bioscience) showed appropriate amplification in real-time PCR reactions [17].

Despite the RNA produced with appropriate concentrations and purity degrees, the samples obtained by the Trizol method showed no amplification in real-time PCR reactions. It corroborates the results found by Witchell et al. (2008) [5], but it did not confirm the data obtained by Körbler et al. (2003) [8] and Antica et al. (2010) [20].

After these results were published, it was decided to check the impact of the residuals and potential contaminants. Thus, the purification step using alcohol in a porous column of the QIAamp®Viral RNA Mini Kit for commercial extraction (Qiagen) was adopted. In brief, after isopropyl alcohol was added to the samples, DNA was filtered using the purification column, according to the manufacturer's recommendations. It was done by transferring the samples to the Kit's purification columns, which were centrifuged at 8,000 g for 1 min. Subsequently, the column products were moved to other tubes. Next, 500μ L buffer AW1 was added to column and it was once again centrifuged at 8,000 g for 1 min. After the filtrate was discarded and the column transferred from the column to another tube, 500 μL Buffer AW2 was added to the solution and a new centrifugation was performed at 14,000 g for 3 min. Then, the column was shifted into a sterile 1.5 mL tube and 60μ L elution buffer was added to it. After the reaction was incubated for 5 min at RT, the tubes were centrifuged at 8,000 g for 1 min and the obtained RNA was stored at –80°C.

A significant improvement was found in the quality of the samples as well as in their adequate amplification in real-time PCR reactions. Thus, RNA extraction from FFPE samples of the three tested protocols became feasible. However, protocol 1 (Ambion commercial Kit) was used as the standard method in current research by taking into account the protocol's practicality and cost.

5. DNA extraction from FFPE tissues

Two different RNA extraction protocols were tested (unpublished data) as described below.

Protocol 1: Phenol-chloroform method [21]

After the sections were prepared and the button cell was completely dried as described in Section 3, tissue digestion was performed by adding 480 μL Tris-EDTA buffer (TE) and 20 μL proteinase K (200 mg/ml) to each sample. Next, these samples were incubated at 37°C for approximately 16 h in the digital thermomixer (Eppendorf AG, Hamburg, Germany). The temperature was then raised up to 90°C for 10 min to inactivate the proteinase K.

One (1) ml of the mixture was added to phenol:chloroform:isoamyl alcohol (Invitrogen Corporation, Carlsbad, CA, USA) at the ratio 25:24:1, respectively. After homogenization using the Vortex Genie 2T (Scientific Industries, Inc., Bohemia, NY, USA), samples were centrifuged at 13,000 g for 15 min at the 5418-R microcentrifuge (Eppendorf AG, Hamburg, Germany). The supernatant was transferred to a fresh 1.5 ml DNase- and RNase-free microfuge tube, and 1 mL of the phenol:chloroform:isoamyl alcohol (25:24:1) mixture was added to it. Then, the centrifugation process was repeated under the same conditions.

The supernatant was transferred to a new 1.5 mL DNase- and RNase-free microtube and 20 μL of 3M sodium acetate and 900 μL of absolute ethanol (Merck KGaA, Darmstadt, Germany) were added to it. The samples were homogenized by inversion and incubated at 20°C for at least 30 min. The samples were centrifuged at 13,000 g for 15 min. The supernatant was discarded and the pellet was completely dried. Samples were resuspended in 50 mL TE buffer and stored at –20°C, until they were used.

Protocol 2: Phenol-chloroform method (modified by our team)

By following the same reasoning used to modify the RNA extraction protocol by Trizol method, the current study made the option to add a DNA purification step using the QIAamp DNA Blood Mini Kit commercial columns (Qiagen, Hilden, Germany) to check the impact from residual and potential contaminants.

As it was described in the previous protocol, the samples were transferred to the column commercial kit – QIAamp DNA Blood Mini Kit (Qiagen, Hilden, Germany), after being incubated with 3M sodium acetate and absolute ethanol. Subsequently, they were centrifuged at 8,000 rpm for 1 min in the 5418- R microfuge (Eppendorf AG, Hamburg, Germany). The filtrate was discarded, and the column was transferred to a new tube and 500 μL of buffer AW1 were added to it. Then, the samples were centrifuged at 8,000 rpm for 1 min in the 5418-R microcentrifuge (Eppendorf AG, Hamburg, Germany).

After the filtrate was discarded and the column transferred to a new tube, a second wash was performed using 500 μL buffer AW2. The samples were centrifuged at 14,000 rpm for 3 min in the 5418-R microcentrifuge (Eppendorf AG, Hamburg, Germany).

Columns were transferred to a 1.5 mL DNases- and RNases-free microtube and 200 μL of buffer AE were added to the center of the column. The samples were then incubated at room temperature for 5 min, and it was followed by centrifugation at 8,000 rpm for 1 min in the 5418-R microfuge (Eppendorf AG, Hamburg, Germany). The samples were stored at –20°C until they were used.

5.1. DNA quality quantification and analysis

The DNA samples' concentration and purity were set by spectrophotometry in NanoDrop®ND-2000 machine (Thermo Fisher Scientific, Wilmington, DE). Samples with absorbance ratios (A280 / A260nm) between 1.7 and 1.9 were considered to be appropriate.

5.2. Results and comments on the herein described DNA extraction protocols

Despite the DNA produced with suitable concentrations and purity degrees, not all the samples obtained by the phenol-chloroform method showed amplification in real-time PCR reactions. It corroborated the results by Witchell et al. (2008) [5], but it did not confirm the data obtained by Körbler et al. (2003) [8] and Antica et al. (2010) [19].

A significant improvement was observed in the quality of the samples and adequate amplification in real-time PCR reactions performed according to method 2. Thus, DNA extraction from FFPE samples of the three tested protocols has become more viable. However, taking into account both the protocol practicality and the cost, the protocol 1 (Ambion commercial Kit) was used as the standard method for the current research.

http://dx.doi.org/10.5772/61581

Acknowledgements

The authors thank Fundação de Amparo à Pesquisa do *Estado de São Paulo* (State of São Paulo Research Support Foundation) for the granted funding.

Author details

Gisele R. Gouveia[1*], Suzete C. Ferreira[2], Sheila A. C. Siqueira[3] and Juliana Pereira[1]

*Address all correspondence to: gisele.rgouveia@gmail.com; gisele.gouveia@usp.br

1 Medical School of University of São Paulo (FMUSP), São Paulo, SP, Brazil

2 Molecular Biology Department of São Paulo Blood Center/Fundação Pró-Sangue, São Paulo, SP, Brazil

3 Pathology Service at Hospital das Clínicas (HC-FMUSP), São Paulo, SP, Brazil

References

[1] Li J, et al. Improved RNA quality and TaqMan®Preamplification method (PreAmp) to enhance expression analysis from formalin fixed paraffin embedded (FFPE) materials. BMC Biotechnol 2008;8:1–11.

[2] Lehmann U, Kreipe H. Real-time PCR analysis of DNA and RNA extracted from formalin-fixed and paraffin embedded biopsies. Methods 2001;25:409–18.

[3] Doleshal M, et al. Evaluation and validation of RNA extraction methods for MicroRNA expression analyses in formalin-fixed, paraffin-embedded tissues. J Molecul Diag 2008;10(3):203–11.

[4] Ribeiro-Silva A, Garcia SB. Estudo comparativo de três diferentes procedimentos para extração de RNA a partir de amostras fixadas em parafina e embebidas em parafina. J Brasil Patol Med Lab 2008;44(2):123–30.

[5] Witchell J, et al. RNA isolation and quantitative PCR from HOPE- and formalin-fixed bovine lymph node tissues. Pathol Res Practice 2008;204:105–11.

[6] Impraim CC, Saiki RK, Erlich HA, Treplitz RL. Analysis of DNA extraction from formalin-fixed, paraffin-embedded tissues by enzymatic and hybridization with sequence-specific oligonucleotides. Biochem Biophys Res Com 1987;142:710–6.

[7] Shibata DK, Amheim N, Martin WJ. Detection of human papiloma virus in paraffin-embedded tissue using the polymerase chain reaction. J Exp Med 1988;167(1):225–30.

[8] Körbler T, et al. A simple method for RNA isolation from formalin-fixed and paraffin-embedded lymphatic tissues. Experiment Molecul Pathol 2003;74:336–40.

[9] Scorsato AP, Telles JEQ. Fatores que interferem na qualidade do DNA extraído de amostras biológicas armazenadas em blocos de parafina. J Brasil Patol Med Lab 2011;47(5):541–8.

[10] Srinivasan M, Sedmak D, Jweell S. Effect of fixatives and tissue processing on the content and integrity of nucleic acids. AJP 2002;161(6):1961–71.

[11] Hamatani K, et al. Improved RT-PCR amplification for molecular analyses with long-term preserved formalin-fixed, paraffin-embedded tissues specimens. J Histochem Cytochem 2006;54(7):773–80.

[12] Coombs NJ, Gough AC, Primrose JN. Optimisation of DNA and RNA extraction from archival formalin-fixed tissue. Nucleic Acids Res 1999;27(16):e12.

[13] Mesquita RA, et al. Avaliação de três métodos de extração de DNA de material parafinado para amplificação de DNA genômico pela técnica da PCR. Pesq Odontol Bras. 2001;15(4):314–9.

[14] Cao W, et al. Comparison of methods for DNA extraction from paraffin-embedded tissues and buccal cells. Cancer Detect Prevent. 2003;27:397–404.

[15] Rivero ERC, et al. Simple salting-out method for DNA extraction from formalin-fixed paraffin-embedded tissues. Pathol Res Practice 2006;202:523–9.

[16] Simonato LE, et al. Avaliação de dois métodos de extração de DNA de material parafinado para amplificação em PCR. J Bras Patol Med Lab 2007;43(2):121–7.

[17] Gouveia GR, Ferreira SC, Sabino EC, Siqueira SAC, Pereira J. Comparação de três protocolos distintos para extração de RNA de amostras fixadas em formalina e emblocadas em parafina. J Brasil Patol Med Lab 2011;47(6):649–54.

[18] Gouveia GR, Ferreira SC, Ferreira JE, Siqueira SAC, Pereira J. Comparison of two methods of RNA extraction from formalin-fixed paraffin-embedded tissue specimens. BioMed Res Int 2014;2014:1–5.

[19] Antica M, Paradzik M, Novak S, Dzebro S, Dominis M. Gene expression in formalin-fixed paraffin-embedded lymph nodes. J Immunol Methods 2010;359(1–2):42–6.

[20] Chomczynski P, Sacchi N. Single step method of RNA isolation by acidic guanidium thiocyanatephenol-chlorophorm extraction. Anal Biochem 1988;162:156–9.

[21] Carvalho VC, Ricci G, Affonso R. Guia de Práticas em Biologia Molecular. São Paulo: Yendis, 2010.

Chapter 3

Nucleic Acid Detection of Major Foodborne Viral Pathogens: Human Noroviruses and Hepatitis A Virus

Haifeng Chen

Additional information is available at the end of the chapter

http://dx.doi.org/10.5772/62112

Abstract

Human noroviruses (hNoVs) and hepatitis A virus (HAV) are the leading cause of foodborne viral illness, and they exact a considerable economic and health toll worldwide. The detection of viral contamination in foods requires highly sensitive and accurate methods, due to the intrinsically low amounts of viruses and the complexity of the sample matrices. In recent years, a wide variety of nucleic acid-based molecular methods have been developed for the detection of hNoVs and HAV, displaying superior sensitivity, specificity, and speed. This chapter aims to provide a summary of the application of the molecular methods for the detection of the two important foodborne viruses.

Keywords: Human noroviruses, hepatitis A virus, foodborne viruses, nucleic acid-based method, detection

1. Introduction

Enteric viruses have been increasingly recognized as the leading causative pathogens in foodborne disease outbreak, causing 66.6% of foodborne illnesses in the United States, compared with 14.2% and 9.7% for *campylobacter* and *salmonella*, respectively [1]. A multitude of foodborne viral pathogens include (but are not limited to) human noroviruses (hNoVs), rotavirus, hepatitis A virus (HAV), hepatitis E virus, astrovirus, aichivirus, sapovirus, parvovirus, enterovirus, and adenovirus [2]. Foodborne viruses are transmitted not only through contaminated food and water, but also in combination with close contact with infected individuals, aerosol contamination of projectile vomit, or through contamination of environmental surfaces. Potentially fecal-contaminated food, such as bivalve molluscan shellfish harvested in polluted water areas, fresh produce irrigated with contaminated water or harvested by an infected worker, and ready-to-eat foods prepared by an infected food handler are a means of infection [3—5]. Of all the foodborne viruses, hNoVs and HAV are the most important foodborne viral pathogens with regard to the severity of the associated illnesses and frequent occurrence worldwide [2]. Both hNoVs and HAV display high environmental stability on contaminated objects, are abundantly excreted in human feces (e.g., exceeding 10^7 viral particles per gram of stool), and have a low infectious dose (1 to 100 infectious viral particles) [2], all of which contribute to the ease of transmission of the viruses within a community. It is commonly noted that one of the most efficient ways to prevent and control the foodborne viral infections is to implement a reliable surveillance system using rapid, sensitive, and precise diagnostics to identify the associated pathogens. Human NoVs do not replicate in cell culture. Wild-type HAV strains are not readily cultivated *in vitro* and the detection is impaired by their slow and inefficient growth in cell culture and lack of apparent cytopathic effect. Cell culture-based systems for determining virus infectivity are currently not available for hNoVs and wild-type HAV. Traditional diagnosis of these foodborne viral pathogens has been reliant on electron microscopy and immunological tests, but these methods lack sufficient sensitivity. While they may be useful for the detection of the viruses in clinical specimens that contain high amounts of viruses, for foods, which harbor potentially small quantities of viruses and may yet cause illness, it is not feasible to use these traditional laboratory methods to detect the viruses. This has led to the development of new, more sensitive and robust detection methods. In recent years, the majority of newly developed detection approaches are nucleic acid-oriented. Nucleic acid-based molecular methods have demonstrated a large improvement in speed, sensitivity, and accuracy of the detection of hNoV and HAV, bringing new insights into the etiology and diagnosis of foodborne viral disease. This chapter will touch upon a number of nucleic acid-based methods that have been developed and applied to detect the two epidemiologically important foodborne viruses.

2. Key notes from norovirus and hepatitis A virus

2.1. Norovirus genome and molecular diversity

Although "winter-vomiting disease" was described in 1929 [6], the responsible viral agent was not discovered until 1972 by Kapikian [7] from fecal materials derived from an outbreak of gastroenteritis among elementary school children in 1968 in Norwalk, Ohio. The virus was named Norwalk virus and was designated as the prototype strain for the group of viruses now called Noroviruses; in the literature, they were previously referred to as small round-structured viruses (SRSVs) by their surface morphology or Norwalk-like viruses. The Norwalk agent was the first enteric virus identified that specifically caused acute gastroenteritis in humans. Successful cloning and sequencing of the NoV genomes have led to progress in understanding viral genome organization and classification. Noroviruses are a member of the family *Caliciviridae* whose name is derived from the Greek word *Calyx* for cup [8]. The viral genome is composed of a single strand of polyadenylated positive-sense, non-enveloped RNA size of approximately 7.6 kb. The linear RNA is organized into three open reading frames (ORFs), designated ORF1, ORF2, and ORF3. ORF1 encodes a 194-kDa protein that is cleaved by the viral cysteine protease into six non-structural proteins including p48, NTPase, p22, Vpg,

http://dx.doi.org/10.5772/62112

protease, and RNA-dependent RNA polymerase. ORF2 codes for the major capsid protein VP1, which folds in two major domains: a shell (S) and a protruding (P) domain. The P domain is comprised of P1 and P2 subdomains. P2 is a highly variable region that is thought to be involved in the binding of the histo-blood group antigens (HBGAs), which are regarded as receptors and host-susceptibility factors for human infection [8, 9].

According to the literature, NoVs have been genetically segregated into at least five genogroups (GI, GII, GIII, GIV, and GV) based on the complete amino acid sequences of the major capsid protein of 164 NoV strains [10]. New genogroups consisting of canine NoVs have also been proposed [11-12]. GI, GII, and GIV are known to cause gastroenteritis in humans. GI and GII contain the majority of human strains, with GV viruses regarded as uncommon human pathogens. GIII viruses were identified in cow, while GV viruses infect mice. Due to their enormous genetic diversity, the viruses within genogroups can be further classified into genetic clusters or genotypes that are defined as containing 14—44% VP1 amino acid sequence difference, where strains have 0—14% difference [10]. Accordingly, there are currently nine recognized genogroup I clusters and 22 genogroup II clusters. Despite their great molecular diversity, Genogroup II, genotype 4 (GII.4) variants have been responsible for the majority of outbreaks and cases in recent years, particularly those associated with person-to-person transmission [13—14]. The genotype GII.4 was first identified to predominate in outbreaks of gastroenteritis in the mid-1990s in countries on five continents [15], and new emerging variants have continued to evolve since then and have become the etiological agents for each of the four global gastroenteritis epidemics [16]. Although the majority of reported NoV outbreaks and cases are derived from person-to-person transmission, it is estimated that approximately14% of them are attributed to food, and 37% of the foodborne outbreaks are caused by mixtures of GII.4 and other genotypes, 10% by all genotype GII.4, and 27% by all other single genotypes [17].

2.2. HAV genome and genotypes

HAV, first identified in 1973 by electron microscopy, is the most common cause of infectious hepatitis with annually causing about 1.4 million clinical cases and 200 million asymptomatic carriers worldwide [18]. HAV is one of the most frequent causes of foodborne viral infection. In the United States, it is estimated that approximately 270,000 people become annually infected with hepatitis A, and most of the infection cases are not reported to health authorities [19]. Epidemics associated with contaminated food or water can occur involving hundreds of thousands of people, such as the epidemic in Shanghai, China in 1988 affected almost 300,000 people due to the consumption of HAV-contaminated clams [20]. Like other enteric viruses, HAV is resilient to environmental stressors. The virus is able to retain infectivity in acidic environments below pH3, and after refrigeration and freezing. HAV is a non-enveloped positive single-stranded RNA virus with a genome of approximately 7.5 kb in length. The virus is classified within the genus *Hepatovirus* of the family of the *Picornaviridae* [21]. The viral genome consists of (i) a 5′-untranslated region (5′-UTR) of about 735 nucleotides; (ii) a single open reading frame (ORF) that is organized into three functional regions termed P1, P2, and P3; (iii) 3′-untranslated region (3′-UTR) with a polyadenylated A tail [22]. The P region encodes the viral capsid polypeptides VP1—VP4, and the P2 and P3 regions encode the non-structural protein. The 5′-UTR is the most conserved region of the genome and therefore is favored for primer design in polymerase chain reaction (PCR) to detect most genotypes. HAV displays a high level of antigenic conservation throughout the viral genome. An immunological study identified the existence of a single human serotype of HAV [23], but ample genetic diversity still exists to classify HAV into six genotypes based upon differences of a 186-bp nucleotide sequence in the VP1—P2A junction region [24—26]. Genotype I, II, and III are associated with human infection, while genotype IV, V, and VI are found in simians [27]. A genotype VII, designated SLF88, was proposed in an earlier study [24], but further analysis of the complete genome and capsid region of additional strains indicated that the genotype VII should be reclassified as genotype IIB [25, 26]. Genotype I and III can be each further divided into subgenotypes IA, IB, IIIA, and IIIB. Genotype I that comprises 80% human HAV strains studied is remarkably prevalent around the globe; subgenotype IA is more common than IB [24]. Since genotype I predominates worldwide, genotyping alone is rarely used to determine the source of a chain of HAV transmissions or outbreaks. Genotype III includes most of the remaining human HAV strains. Genotype II contains two subgenotypes: IIA and IIB.

3. Nucleic acid-based detection

3.1. Direct nucleic acid probe hybridization

Direct nucleic acid probe hybridization was the first molecular technique developed for the detection of suite of enteric viral pathogens. This technique can be used in diagnostics in several major formats: solid-phase, solution or liquid-phase, and *in situ* hybridization. In hybridization assays, oligonucleotide probes (single-stranded RNA or cDNA) that are complementary to the target genomic sequence of interest were labeled with signal reporters, which include radioactive molecules, chemiluminescence, or fluorescent agents. After hybridization, the probe signal from the reporter can be visualized via radioactivity, fluorescence, or color development. Detection of the probe signal indicates the presences of nucleotide sequences of interest that have high sequence similarities to the probe. HAV has been detected using these techniques such as dot blot hybridization [28—30] and *in situ* hybridization [31]. Dot blot hybridization assays were also used for detection of Norwalk viruses in 55 stool specimens from human volunteers with 27 samples tested positive [32]. A potential advantage of the direct hybridization technique is the low cost of the assay and decreased risk of cross-contamination [33]. However, the disadvantage is that the detection sensitivity is often low (approximately 10^4 virus particles) [28], thus limiting its practical application in detecting low numbers of viruses in clinical specimens, food, and environmental samples.

3.2. Nucleic acid amplification

Despite a number of reports describing the use of the direct probe hybridization technique, new molecular detection methods that incorporate the amplification of target nucleic acids are now being developed and predominantly used for the detection of foodborne viral pathogens in samples of different origins. Nucleic acid amplification offers an edge over direct probe hybridization by enhancing the detection sensitivity through amplifying target nucleic acids extracted from samples.

3.2.1. Reverse transcription-PCR

Since the first demonstration of the PCR process in 1985[34], this technology has been widely used for the detection of foodborne pathogens. As RNA cannot be directly used as a template for PCR amplification, reverse transcription-PCR (RT-PCR) is employed for the amplification of viral RNA. RT-PCR can generally be carried out either in a one-step or two-step format. One-step RT-PCR combines the first-strand cDNA synthesis reaction (reverse transcription) and PCR amplification in a single tube, minimizing reaction setup and risk of carryover contamination. Alternatively, the two-step RT-PCR starts with the reverse transcription of either total RNA or poly (A) RNA into cDNA using a combination of sequence-specific primers, oligo (dT) or random primers in the presence of reverse transcriptase. The resulting cDNA then serves as a template for the initiation of PCR amplification in a separate tube. Separation of reverse transcription and PCR processes allows greater flexibility when choosing primers and polymerase than the one-step RT-PCR system, which allows for the use of sequence-specific primers only.

The first RT-PCR assays for detecting NoVs were described within two years of the successful cloning and sequencing of the Norwalk virus genome in the early 1990s [32, 35]. Application of this technology has allowed the detection of NoVs from samples of different origins and has generated a great deal of sequence information on various NoV strains. The sensitivity and specificity of RT-PCR assays are strongly associated with primer design. RT-PCR tests to detect NoVs are challenged by the high molecular diversity of the viruses since new variant strains continue to evolve incessantly [36]. It is difficult to select a single oligonucleotide primer set with sufficient sensitivity and specificity to detect all the NoV strains [33]. Different primer sets targeting multiple regions of the viral genome have been designed and evaluated in RT-PCR assays. A highly conserved RNA-dependent RNA polymerase region has been favored for primer design and amplification ([37—39]. Other regions such as capsid region, 2C helicase and ORF3 regions have also been targeted for amplification [40—43]. In addition, RT-PCR assays using different primer combinations in nested (two primers) or semi-nested (one primer) format have been performed to increase the likelihood of NoV detection [44, 45]. It has been reported that higher detection sensitivity (10 to 1,000 times more sensitive than single round RT-PCR) has been achieved by implementation of this strategy [44].

HAV was one of the first enteric viruses for which RT-PCR assays have been developed [46]. In contrast to NoVs, many human HAV strains across different genotypes can be amplified using a single pair of primers targeting genes coded for structural proteins, e.g., VP1—2A and VP3—VP1 junction regions [47, 48]. 5′-untranslated region (5′UTR) primers were also used in RT-PCR assays to detect HAV from clinical and environmental samples [48].

After amplifying a target of interest, post-amplification analysis is necessary to interpret the results. The simplest method is to run the amplified products on ethidium bromide- or SYBR Green dye-stained agarose gels. A band that is of the expected size to that of the positive control and/or molecular weight markers is considered a positive result. However, this method does not provide additional reassurance as to the specificity of the amplification. Analysis of the RT-PCR products by restriction fragment length polymorphism (RFLP) could discriminate genetic variants of HAV of different origins; this resolution has been enhanced by combining with the results of single-strand confirmation polymorphism (SSCP) analysis [49]. PCR-SSCP has been considered as a rapid and cost-effective approach to examining genetic diversities among hNoV strains [50] as well as HAV [51—53]. Hybridization assays such as dot blot and Southern blot have been used to identify and confirm NoV- and HAV-specific amplicons [32, 54—57]. The reverse line blot hybridization method was utilized to detect NoV RT-PCR products and to genotype the virus strains [58]. DNA microarray technology has been used to analyze amplified products for detecting and genotyping hNOVs and HAV discussed in the section below. Direct sequencing of amplified products provides detailed molecular information not only for confirming the specificity of the amplicons but also for classifying or subtyping virus strains [59].

3.2.2. *Quantitative real time RT-PCR (RT-qPCR)*

A RT-qPCR system combines amplification of target nucleic acids with amplicon detection in the same reaction tube, eliminating the necessity of further post-amplification analysis. RT-qPCR is currently the method of choice for the detection of hNoVs and HAV in many molecular diagnostic laboratories. This method has become the gold standard for quantification of viral load based on a reference standard curve. Two principle approaches are commonly utilized for the detection of enteric viruses in RT-qPCR assays: DNA-binding fluorogenic dyes and sequence-specific oligonucleotide probes. Selected application of these approaches for detection of NoV and HAV is outlined in Tables 1 and 2.

SYBR Green is commonly used as a dye for the quantification of double-stranded DNA in qPCR methodology. The binding of SYBR Green is non-specific; it binds indiscriminately to any double-stranded DNA including non-specific amplification and primer-dimers. To distinguish virus-specific amplified products from non-specific primer-dimers, a melting curve analysis is generally used as the virus-specific products have a higher dissociation temperature. A number of SYBR Green-based RT-qPCR assays have been developed for the detection of hNoVs by targeting different genomic regions such as the capsid region [60, 61] and RNA polymerase [61—64]. SYBR Green-based RT-qPCR employing 5′—UTR region primers was used to detect as low as 5 Tissue Culture Infectious Dose 50% ($TCID_{50}$) per gram in seeded oyster samples [65]. Using SYBR Green RT-qPCR with VP3—VP1 junction primers, HAV could be detected in all eight tested ocean water samples with viral loads varying from 90 to 3523 HAV copies/L near the mouth of Tijuana River, and 347 to 2656 copies/L near the Imperial Beach pier in San Diego [66].

Approach	Genogroup	Target Region	Lmit of Detection	Detection/ Quantitation	Sample tested	References
SYBR Green	Not reported	RNA polymerase	Not reported	D	Stool	[62]
	GI	Capsid	1 RT-PCR unit	Q	Stool	[60]
	Not reported	RNA polymerase	Up to 5 logs	D	Stool	[63]
	GI/GII	Capsid in GI/ RNA polymerase in GII	25,000 RNA copies/gram	D	Stool	[61]
	Not reported	RNA polymerase	6 log GII cDNA copies/100ml	Q	Enviromental water	[64]
TaqMan probe	GI/GII	ORF1-ORF2 junction	2.0x10^4RNA	Q	Stool	[67]
	GI/GII/GIV	ORF1-ORF2 junction	<10GII, <100GI RNA copies	Q	Stool/water	[70]
	GI/GII	Capsid	50 GII copies, 500 GI copies	Q	Stool	[68]
	GI/GII/GIV	ORF1-ORF2 junction	16.9GI, 6.3GII, 43GIV copies	Q	Stool, vomitus, anal swab	[73]
	GI/GII	ORF1-ORF2 junction	23GI, 33GII copies	D	Shellfish & springwater	[83]
	GI/GII	Capsid	3~7 RT-PCR unit	Q	Strawberry	[72]
Hybridization probe	GI/GII	ORF1-ORF2	50 copies	D	Stool	[79]
	GI/GII	ORF1-ORF2	1~10GI, 1~100GII RNA copies	D	Stool	[78]

Table 1. Selected RT-qPCR assays for detection of human noroviruses

There are two major groups of sequence-specific oligonucleotide probes: hydrolysis (e.g., TaqMan) probes and hybridization probes (e.g., molecular beacons and fluorescence resonance energy transfer probes); both groups are homologous to the internal region of amplified products. TaqMan probes have been frequently used in RT-qPCR for detecting hNOVs [67–73]. While RNA polymerase and capsid genes are the primary targets for amplification, within the NoV genomes, a junction of ORF1—ORF2—polymerase—capsid has been demonstrated to be the most highly conserved region that can serve as an effective target for amplification.

Approach	Genotype	Target Region	Lmit of Detection	Detection/ Quantitation	Sample tested	References
SYBR Green	Not reported	VP3-VP1	2-4 log copies/l	D	Ocean water	[66]
	Not reported	5′-UTR	5 $TCID_{50}$/g	Q	Oyster	[65]
TaqMan probe	IA, IB, IIA, IIIA	5′-UTR, VP1-2A, VP3, VP1, VP2-VP3	50 copies for IIB, 500 for IA, IB, IIA and IIIB	Q	Stool, serum	[77]
	IA, IB	5′-UTR	10 copies or 0.05	Q	Shellfish, stool	[75]
	IA, IB, IIA, IIB, IIIA	5′-UTR	40 copies or 0.5	D	Cell culture	[74]
	IB	5′-UTR	491 copies	Q	Shellfish, springwater	[83]
	IB	5′-UTR	0.2 PFU, 63PFU	D	Cell culture, green onion	[76]
Molecular beacons	IB	5′-UTR	20 PFU	D	Seeded groundwater	[80]

Table 2. Selected RT-qPCR assays of detection of hepatitis A virus

Kageyama and colleagues were the first to describe the junction-targeting RT-qPCR assay [67]. Their studies showed a better detection rate in 80 of 81 (99%) stool samples positive by electron microscopy, compared to conventional RT-PCR assays that detected 77% when targeting the polymerase and 83% when targeting the capsid N/S regions, respectively, in the same panel of stool specimens [67]. In an effort to design assays capable of detecting all genogroups of these highly diverse viruses, this ORF1—ORF2 junction region has become the most widely used for primer and probe design in RT-qPCR tests. A TaqMan RT-qPCR assay using a probe for the 5′-UTR of HAV genome was able to detect 40 copies of RNA transcripts and 0.5 infectious units (IU) in cell culture strains and clinical fecal specimens, respectively [74]. Constafreda et al. (2006) also developed a TaqMan RT-qPCR method targeting 5′-UTR for quantitative detection of HAV from clinical specimens (stool and serum) and shellfish samples, and the detection limit was 0.05 IU or 10 copies of single-stranded RNA transcripts [75]. Nested RT-PCR assays combining conventional PCR, nested PCR and qPCR have been used to detect as low as 0.02 plaque forming units (PFU) of HAV from cell culture and 63 PFU from green onions [76]. Six subtype-specific RT-qPCR assays using hydrolysis probes were developed for HAV detection and subtyping [77], with limit of detection at 50 genome copies/assay for subtype IIB, 500 genome copies/assay for IA, IB, IIA, and IIIB, and 5,000 genome copies/assay for IIIA. Thirty-five clinical stool and serum specimens were tested with this method. Only a single discrepant result was observed for a serum specimen provided as IB subtype by VP1/2A region sequencing and identified as IA by the subtype-specific RT-qPCR assays.

Hybridization probes including molecular beacons and fluorescence resonance energy transfer (FRET) probes have been used in RT-qPCR assays for detecting NoVs [78, 79]. Abd El Galil et al. showed that a molecular beacon RT-qPCR assay targeting the HAV highly conserved 5′-UTR region could detect as lower as 20 PFU HAV in seeded groundwater samples in combination with immunomagnetic separation [80]. Molecular beacon probes have also been used in another target amplification method: nucleic acid sequence-based amplification described below.

3.2.3. Multiplex PCR

Multiplex PCR-based methods are designed to simultaneously amplify more than one target nucleic acids using different sets of primers in a single reaction. Primers of different targets should not be complementary to each target and can work efficiently at the same annealing temperature during PCR.

A multiplex RT-PCR method was developed for simultaneous detection of HAV, norovirus, and poliovirus type (PV1) using three different sets of primers with detection limits of ≤ 1 infectious unit (HAV and PV1) and 1 RT-PCR unit (NoV) [81]. Noroviruses (GI and GII) and HAV from different food matrices such as lettuce, strawberry, green onions, and bivalve mollusks were detected using a multiplex RT-qPCR assay [82, 83]. A multiplex RT-qPCR for simultaneous detection and quantification of GI, GII, and GIV noroviruses was recently reported [84]. Several multiplex nucleic acid diagnostic platforms for simultaneous detection of a range of pathogenic enteric pathogens including hNoVs are now commercially available [85]. Although multiplex PCR allows rapid and cost-effective detection of several targets (viruses) in a single reaction, different targets may compete with each other for resources such that a highly abundant target would get more chances to be amplified, thus preventing less abundant ones from getting detected. In addition, multiplex PCR tend to have decreased sensitivity as compared to standard single PCR.

3.2.4. Digital PCR

Digital PCR has recently been described as a novel approach to nucleic acid detection and quantification. It is a different method of absolute quantification relative to conventional quantitative PCR, because it directly counts the number of target nucleic acid molecules rather than relying on reference standards. This improvement is achieved by partitioning sample amplification reactions into tens to thousands of picoliter-scale compartments on microfluidic chips or microdroplets so that each mini-reaction contains zero or one copy of the target nucleic acid molecule. A comparative study of digital RT-PCR (RT-dPCR) and RT-qPCR for the detection and quantification of HAV and NoVs in lettuce and water samples was recently reported [86]. This RT-dPCR assay showed that the sensitivity was either comparable or slightly (around 1 $\log_{10}$) decreased to that of RT-qPCR for detecting viral RNA and cDNA of HAV and NoV, but the viral recoveries were found to be significantly higher than that of RT-qPCR for NoV GI and HAV in water, and for NoV GII and HAV in lettuce. In addition, this RT-dPCR was more tolerant to inhibitory substances present in lettuce.

3.2.5. Nucleic Acid Sequence-Based Amplification (NASBA)

Nucleic acid sequence-based amplification is an isothermal target amplification process for amplifying RNA. A NASBA reaction consists of reverse transcriptase, T7 RNA polymerase, and RNase-H with two target sequence-specific oligonucleotide primers. A NoV NASBA assay showed equivalent analytical sensitivities with RT-PCR using the NoV GII primer sets described by Kageyama et al [67] but provided less consistent signals [87]. A molecular beacon real-time NASBA method was developed to detect NoV GII from environmental samples, with 88% sensitivity compared to conventional RT-PCR [88]. This NASBA technology has been applied to detect as little as 0.4 ng of HAV RNA/ml using primers targeting the VP1 and VP2 capsid genes [89]. A multiplex NASBA and microtiter plate hybridization system was developed to simultaneously detect HAV and rotavirus where 400 PFU/ml HAV were detected with reduced time and cost compared to monoplex system [90]. Using established primer pairs, multiplex NASBA assays were developed for simultaneous detection of HAV and NoV GI and GII in spiked ready-to-eat foods. All three viruses were simultaneously detected at initial inoculum levels of 10(0) to 10(2) RT-PCR units [91].

3.2.6. Reverse transcription loop-mediated isothermal amplification (RT-LAMP)

RT-LAMP is a one-step non-PCR nucleic acid amplification that is performed at a constant temperature between 60 and 65°C. Unlike NASBA, it requires only two enzymes instead of three, namely, reverse transcriptase and DNA polymerase. A genogroup-specific RT-LAMP assay has been developed using 9 and 13 specially designed primers containing mixed bases for genogroup I (GI) and II (GII), respectively, and showed the limits of detection between 10^2 and 10^3 copies/tube for GI and GII. Compared to conventional RT-PCR, the clinical sensitivity and specificity of the RT-LAMP were 100% and 94% for GI, and 100% and 100% for GII, respectively [92]. Commercial loopamp NoV GI and GII detection kits were evaluated using 510 clinical fecal specimens; the sensitivity of GI (83.3%) was less than that of GII (97.4%) with regard to genogroup-specific RT-qPCR [93]. A single tube, real-time HAV RT-LAMP assay using seven primer sets was applied to identify three different subgenotypes of HVA (IA, IB, and IIIB) with detection limits of 0.4—0.8 focus forming units *per* reaction [94].

3.3. Nucleic acid amplification coupled with DNA microarray detection

DNA microarray technology consists of numerous individual target-probe hybridization reactions that are performed in a single assay. The intrinsic ability of this technology to simultaneously analyze thousands of specific DNA sequences presents a significant advantage in parallel identification of a broad spectrum of microbial pathogens. Additionally, a DNA microarray composed of well-designed probes has the potential to discover novel viruses or pathogens not well-represented in the current sequence database. Not surprisingly, this technology, coupled with virus-specific monoplex or multiplex RT- PCR amplification, has enabled sensitive detection and identification of a number of enteric viruses including hNoVs and HAV from clinical specimens, environmental samples, and virus-infected cell cultures [95—100]. However, unbiased amplification with virus-specific PCR is often complicated by the existence of enormous genetic diversities in foodborne viruses. It would be advantageous to develop amplification approaches that do not rely on specific pathogen sequence information, yet can produce sufficient target nucleic acids from minute amounts of starting materials of viral, bacterial, plant, and animal origins for microarray analysis [101]. Recently, a sequence-independent isothermal RNA amplification approach has been developed to amplify various enteric viral nucleic acids of hNoVs, HAV, and coxsackievirus for microarray analysis [102]. Utility of this microarray platform and amplification strategy allows not only discerning genotypic information on hNoVs but also detection of mixed viral agents (hNoVs and HAV) present in the same fecal specimen [103]. Microarray detection of random-primed PCR products from a range of gastrointestinal viruses including Norwalk virus was reported recently [104].

3.4. Detection of infectious viruses

As mentioned above, there are no efficient cell culture systems available for hNoV propagation, and wild-type HAV strains are difficult to grow *in vitro*. In absence of effective culture-based infectivity assays, development of rapid and sensitive molecular methods for reliably detecting infectious vial particles to determine virus infectivity is a key issue for the application of food risk management. Integrity of the virus capsid and its genome are essential for virus infectivity; both have been targeted for the development of methods for predicting virus infectivity. Nuanualsuwan and Cliver (2002) described a method in their effort to correlate RT-PCR data with virus infectivity [105]. In their study, HAV, vaccine poliovirus, and feline calicivirus were inactivated by ultraviolet light, hypochlorite, or heating at 72°C. They observed that the inactivated viruses, which were treated with proteinase K and ribonuclease before RT-PCR, did not yield positive amplicons [106]. Integrated RT-qPCR approaches have been used to discriminate the infectivity status of NoVs based on the assumption that infectious virus particles would more efficiently bind to the appropriate receptors than non-infectious viruses [106, 107]. Long-range RT-qPCR has been used to test the integrity of the NoV genome following 72°C heat treatment [108]. Recent studies on the use of nucleic acid intercalating dyes such as ethidium monoazide (EMA) and propidium monoazide (PMA) in conjunction with RT-PCR or RT-qPCR to distinguish between infectious and non-infectious enteric viruses including hNoVs and HAV have been reported [109—111]. However, PMA RT-PCR could not differentiate infectious Norwalk virus from non-infectious Norwalk virus, although it was able to differentiate selectively between infectious and noninfectious murine NoV, coming to the conclusion that PMA RT-PCR can be used to detect intact, potentially infectious viruses only under specified conditions [108, 110]. A real-time NASBA combined with enzymatic treatment of proteinase K and RNase has been developed to discriminate the infectious from the heat-inactivated hNoVs [112].

4. Concluding remarks

Over the past few years, nucleic acid-based molecular methods have been developed, refined, and used for the detection of hNoVs and HAV that are most commonly associated with the transmission of foodborne viral disease. In particular, RT-qPCR has emerged as a preferred method for the detection and quantification of the viruses due to its high sensitivity, reproducibility, speed, and minimization of risk of carry-over contamination. Recent advances in high-throughput next-generation sequencing (NGS) technologies have opened new avenues for genomic research and diagnostic applications. It is expected that utilization of NGS in viral metagenomics and whole genome sequencing will highly improve the opportunities for identifying viruses of different origins including those that are too divergent to be detected by PCR or other molecular approaches. However, the identification of hNoVs and HAV in vast ranges of food matrices is still a demanding task that is largely attributed to some factors such as intrinsically low quantities of contaminated viruses and broad chemical composition of food, which may inhibit the activity of enzymes used in the molecular detection. This requires a meticulous investigation into sample preparation procedures to obtain acceptable recovery of viral RNA for downstream analysis. Nucleic acid-based molecular methods have a disadvantage in that the majority of the reported assays for detection of foodborne virus contamination do not have standardized protocols, and can vary from laboratory to laboratory. This could be due to several reasons, such as different approaches required for preparation of viral nucleic acids from different test matrices, and a lack of multi-laboratory validation of promising procedures. There is a need for harmonized standards and quality control of the reagents used. Moreover, in many cases, molecular methodology has focused on merely detecting the presence of viral nucleic acids that is not necessarily associated with the detection of infectious particles, although some studies stated above have shown promising differentiating results. Research on developing new methods to accurately determine the virus infectivity is eventually needed to gauge health risk.

Acknowledgements

The author is grateful to Marianne Solomotis and Michael Kulka for reading this manuscript.

Author details

Haifeng Chen*

Address all correspondence to: haifeng.chen@fda.hhs.gov

U.S. Food and Drug Administration, CFSAN/OARSA/DMB, Laurel, MD, USA

The views and opinions expressed in this article are those of the author and do not necessarily reflect the official views of the U.S. Food and Drug Administration (FDA).

References

[1] Mead PS, Slutsker L, Dietz V et al. Food-related illness and death in the United States. Emerg Infect Dis. 1999;5:607—625.

[2] FAO/WHO [Food and Agriculture Organization of United Nations/World Health Organization]. (2007). Viruses in food: scientific advice to support risk management activities: Meeting Report.Microbiological Risk Assessment Series No. 13. Bilthoven, the Netherlands.

[3] Nainan OV, Xia G, Vaughan G, Margolis HS. Diagnosis of hepatitis a virus infection: a molecular approach. Clin Microbiol Rev. 2006;19:63—79.

[4] Montano-Remacha C, Ricotta L, Alfonsi V et al. Hepatitis A outbreak in Italy, 2013: a matched case-control study. Euro Surveill. 2014;18:19(37).

[5] Hall Aj, Eisenbart VG, Etingue Al et al. Epidemiology of foodborne norovirus outbreaks, United States, 2001-2008. Emerg Infect Dis. 2012;18:10.

[6] Zahorsky J. Hyperemesis biemis or the winter vomiting disease. Arch Pediatr. 1929;46:391—395

[7] Kapikian AZ, Wyatt RG, Dolin R, et al. Visualisation by immune electron microscopy of a 27 nm particle associated with acute infectious nonbacterial gastroenteritis. J Virol. 1972;10:1075—1081.

[8] Glass RI, Parashar UD, and Este MK. Norovirus Gastroenteritis. N Eng J Med 2009;361:1776—1785.

[9] Huang P, Farkas T, Marionneau S, et al. Noroviruses bind to human ABO, Lewis, and secretor histo-blood group antigens: identification of 4 distinct strain-specific patterns. J Infect Dis. 2003 188:19—31.

[10] Zheng DP, Ando T, Fankhauser RL et al. Norovirus classification and proposed strain nomenclature. Virology. 2006;346:312—323.

[11] Martella V, Lorusso E, Decaro N et al. Detection and molecular characterization of a canine norovirus. Emerg Infect Dis. 2008;14:1306–1308.

[12] Tse H, Lau SKP, Chan WM et al. Complete genome sequences of novel canine noroviruses in Hong Kong. J Virol. 2012,86:9531–9532.

[13] Matthews JE, Dickey BW, Miller RD et al. The epidemiology of published norovirus outbreaks: a review of risk factors associated with attack rate and genogroup. Epidemiol Infect. 2012;140:1161—1172.

[14] Vega E, Barclay L, Gregoricus N et al. Genotypic and epidemiologic trends of norovirus outbreaks in the United States, 2009 to 2013. J Clin Microbiol. 2014;52:147—155.

[15] Noel JS, Fankhauser RL, Ando T et al. Identification of a distinct common strain of "Norwalk-like viruses" having a global distribution. J Infect Dis. 1999;179:1334—1344.

[16] Siebenga JJ, Vennema H, Renchens B et al. Epochal evolution of GGII.4 norovirus capsid proteins from 1995 to 2006. J of Virol. 2007; 81:9932—9941.

[17] Verhoef L, Hewitt J, Barclay L et al. Norovirus genotype profiles associated with foodborne transmission, 1999-2012. Emerg Infect Dis. 2015; 21:592—599.

[18] World Health Organization. Health Topics: Hepatitis. 2014. Geneva, Switzerland. Available from: http://www.who.int/topics/hepatitis/en/.

[19] Armstrong GL, Bell BP. Hepatitis A virus infections in the United States: model-based estimates and implications for childhood immunization. Pediatrics. 2002;109:839—845.

[20] Halliday ML, Kang LY, Zhou TK et al. An epidemic of hepatitis A attributable to the ingestion of raw clams in Shangihai, China. J Infect Dis. 1991;164:852—859.

[21] Minor PD. Picornaviridae, classification and nomenclature of viruses. The Fifth Report of the International Committee on Taxonomy of Viruses. Arch Virol. 1991;Suppl. 2:320—326.

[22] Rueckert PR & Wimmer E. Systematic nomenclature of picornavirus protein. J of Virol. 1984; 50: 957—959.

[23] Lemon SM, Binn LN. Antigenic relatedness of two strains of hepatitis A virus determined by cross-neutralization. Infect Immun. 1983;42:418—420.

[24] Robertson B H, Jansen RW, Khanna B et al. Genetic relatedness of hepatitis A virus strains recovered from different geographical regions. J Gen Virol.1992; 73:1365—1377.

[25] Costa-Mattioli M, Di Napoli A, Ferre′ V et al. Genetic variability of hepatitis A virus. J Gen Virol 2003;84:3191—3201.

[26] Lu L, Ching KZ, Salete de Paula V et al. Characterization of the complete genome sequence of genotype II hepatitis A virus (CF53/Berne isolate). J Gen Virol. 2001; 85:2943—2952.

[27] Nainan OV, Margolis HS, Robertson BH et al. Sequence analysis of a new hepatitis A virus naturally infecting cynomolgus macaques (Macaca fascicularis). J Gen Virol. 1991;72:1685—1689.

[28] Jiang X, Estes MK, Metcalf TG et al. Detection of hepatitis A virus in seeded estuarine samples by hybridization with cDNA probes. Appl Environ Microbiol. 1986; 52:711—717.

[29] Ticehurst JR, Feinstone SM, Chestnut T, Tassopoulos NC, Popper H, Purcell RH. Detection of hepatitis A virus by extractionof viral RNA and molecular hybridization. J Clin Microbiol. 1987;25:1822—1929.

[30] Jiang X, Estes MK, Metcalf TG. Detection of hepatitis A virus by hybridization with single-stranded RNA probes. Appl Environ Microbiol. 1987;53:2487—2495.

[31] Jiang X, Estes MK, Metcalf TG. In situ hybridization for quantitative assay of infectious hepatitis A virus. J Clin Microbiol. 1989;27:874—879.

[32] Jiang X, Wang J, Graham DY, Estes MK. Detection of Norwalk virus in stool by polymerase chain reaction. J Clin Microbiol. 1992;30:2529—2534.

[33] Atmar RL and Estes MK. Diagnosis of noncultivatable gastroenteritis viruses, the human caliciviruses. Clin Microbiol Rev. 2001;14:15—37.

[34] Saiki R, Scharf S, Faloona F et al. Enzymatic amplification of beta-globin genomic sequences and restriction site analysis for diagnosis of sickle cell anemia. Science. 1985;230:1350—1354.

[35] De Leon R, Matsui SM, Baric RS et al. Detection of Norwalk virus in stool specimens by reverse transcriptase-polymerase chain reaction and nonradioactive oligoprobes. J Clin Microbiol. 1992;30:3151—3157.

[36] Boon D, Mahar JE, Abente EJ, Kirkwood CD, Purcell RH, Kapikian AZ, Green KY, Bok K. Comparative evolution of GII.3 and GII.4 norovirus over a 31-year period. J Virol. 2011; 85:8656—8666.

[37] Ando T, Monroe SS, Gentsch JR et al. Detection and differentiation of antigenically distinct small round structured viruses (Norwalk-like viruses) by reverse transcription-PCR and Southern hybridization. J Clin Microbiol. 1995;33:64–71.

[38] Green J, Gallimore CI, Norcott JP et al. Broadly reactive reverse transcriptase polymerase chain reaction inthe diagnosis of SRSV-associated gastroenteritis. J Med Virol. 1995;47:392—398.

[39] Le Guyader F, Estes MK, Hardy ME et al. Evaluation of a degenerate primer for the PCR detection of human caliciviruses. Arch Virol. 1996;141:2225—2235.

[40] Green SM, Lambden PR, Caul EO and Clarke IN. Capsid sequence diversity in small round structured viruses from recent UK outbreaks of gastroenteritis. J Med Virol. 1997;52:14—19.

[41] Matsui SM., Kim JP, Greenberg HB et al. The isolation and characterization of a Norwalk virus-specific cDNA. J Clin Investig. 1991;87:1456—1461.

[42] Wang J, Jiang X, Madore HP et al. Sequence diversity of small round structured viruses. J Virol. 1994;68:5982—5990.

[43] Moe CL, Gentsch J, Grohmann G et al. Application of PCR to detect Norwalk virus in fecal specimens from outbreaks of gastroenteritis. J Clin Microbiol. 1994;32:642—648.

[44] Green J, Henshilwood K, Gallimore CI, Brown DW, Lees DN.A nested reverse transcriptase PCR assay for detection of small round-structured viruses in environmentally contaminated molluscan shellfish. Appl Environ Microbiol. 1998;64:858—863.

[45] Stene-Johansen K, Grinde B. Sensitive detection of human Caliciviridae by RT-PCR. J Med Virol. 1996;50:207—213.

[46] Jansen RW, Siegl G, Lemon SM. Molecular epidemiology of human hepatitis A virus defined by an antigen-capture polymerase chain reaction method. Proc Natl Acad Sci U S A. 1990;87:2867—2871.

[47] Hutin YJF, Pool V, Cramer EH et al. A multistate foodborne outbreak of hepatitis A. N Engl J Med. 1999;340:595—602.

[48] Pina S, Buti M, Jardi R et al. Genetic analysis of hepatitis A virus strains recovered from the environment and from patients with acute hepatitis. J Gen Virol. 2001;82:2955—2963.

[49] Goswami BB, Burkhardt W 3rd, Cebula TA. Identification of genetic variants of hepatitis A virus. J Virol Methods. 1997;65:95—103.

[50] Sasaki Y, Kai A, Hayashi Y, Shinkai T et al. Multiple viral infections and genomic divergence among noroviruses during an outbreak of acute gastroenteritis. J Clin Microbiol. 2006;44:790—797.

[51] Sulbaran Y, Gutierrez CR, Marquez B et al. Hepatitis A virus genetic diversity in Venezuela: exclusive circulation of subgenotype IA and evidence of quasispecies distribution in the isolates. J Med Virol. 2010;82:1829—1834.

[52] Mackiewicz V, Roque-Afonso AM, Marchadier E etal. Rapid investigation of hepatitis A virus outbreak by single strand conformation polymorphismanalysis. J Med Virol. 2005;76:271—278.

[53] Fujiwara K, Yokosuka O, Ehata T et al. PCR-SSCP analysis of 5'-nontranslated region of hepatitis A viral RNA: comparison with clinicopathological features of hepatitis A. Dig Dis Sci. 2000;45:2422—2427.

[54] Willcocks MM, Silcock JG, Carter MJ. Detection of Norwalk virus in the UK by the polymerase chain reaction. FEMS Microbiol Lett. 1993;112:7—12.

[55] Kohn MA, Farley TA, Ando T et al. An outbreak of Norwalk virus gastroenteritis associated with eating raw oysters. Implications for maintaining safe oyster beds. JAMA 1995;273:466—471.

[56] Goswami BB, Koch WH, Cebula TA. Detection of hepatitis A virus in Mercenaria mercenaria by coupled reverse transcription and polymerase chain reaction. Appl Environ Microbiol. 1993;59:2765—2770.

[57] Nainan OV, Cromeans TL and Margolis HS. Sequence-specific, single-primer amplification and detection of PCR products for identification of hepatitis viruses. J Virol Methods.1996;61:127—134.

[58] Vinjé J & Koopmans MP. Simultaneous detection and genotyping of "Norwalk-like viruses" by oligonucleotide array in a reverse line blot hybridization format. J Clin Microbiol. 2000;38:2595—2601.

[59] Robertson BH, Khanna B, Nainan OV, Margolis HS. Epidemiologic patterns of wild-type hepatitis A virus determined by genetic variation. J Infect Dis. 1991;163:286—292.

[60] Richards GP, Watson MA, Kingsley DH. A SYBR green, real-time RT-PCR method to detect and quantitate Norwalk virus in stools. J Virol Methods. 2004;116:63–70.

[61] Pang X, Lee B, Chui L, Preiksaitis JK,Monroe SS.. Evaluation and validation of real-time reverse transcription-PCR assay using the LightCycler system for detection and quantitation of norovirus. J Clin Microbiol.2004; 42:4679–4685.

[62] Miller I, Gunson R, Carman WF. Norwalk like virus by light cycler PCR. J Clin Virol. 2002;25:231–232.

[63] Schmid M, Oehme R, Schalasta G, Brockmann S, Kimmig P; Enders G. Fast detection of noroviruses using a real-time PCR assay and automated sample preparation. BMC Infect Dis 2004;4:15.

[64] Laverick MA, Wyn-Jones AP, Carter MJ. Quantitative RT-PCR for the enumeration of noroviruses (Norwalk-like viruses) in water and sewage. Lett Appl Microbiol. 2004;39:127—136.

[65] Casas N, Amarita F, de Marañón IM. Evaluation of an extracting method for the detection of Hepatitis A virus in shellfish by SYBR-Green real-time RT-PCR. Int J Food Microbiol. 2007;120:179—185.

[66] Brooks HA, Gersberg RM, Dhar AK. Detection and quantification of hepatitis A virus in seawater via real-time RT-PCR. J Virol Methods. 2005;127:109—118.

[67] Kageyama T, Kojima S, Shinohara M, Uchida K, Fukushi S, Hoshino FB, Takeda N, Katayama K. Broadly reactive and highlysensitive assay for Norwalk-like viruses based on real-time quantitative reverse transcription-PCR. J Clin Microbiol. 2003;41:1548—1557.

[68] Hohne M, Schreier E. Detection and characterization of norovirus outbreaks in Germany: application of a one-tube RT-PCR using afluorogenic real-time detection system. J Med Virol. 2004;72:312—319.

[69] Dreier J, Stormer M, Made D, Burkhardt S, Kleesiek K. Enhanced reverse transcription-PCR assay for detection of norovirus genogroup I. J Clin Microbiol. 2006; 44:2714—2720.

[70] Trujillo AA, McCaustland KA, Zheng DP, Hadley LA, Vaughn G, Adams SM, Ando T, Glass RI, Monroe SS. Use of TaqMan real-time reverse transcription-PCR for rapid

detection, quantification, and typing of norovirus. J Clin Microbiol. 2006; 44:1405–1412.

[71] Scipioni A, Bourgot I, Mauroy A, Ziant D, Saegerman C, Daube G,Thiry E. Detection and quantification of human and bovine noroviruses by a TaqMan RT-PCR assay with a control for inhibition. Mol Cell Probes. 2008;22:215–222.

[72] Park Y, Cho YH, Jee Y, Ko G. Immunomagnetic separation combined with real-time reverse transcriptase PCR assays for detection of norovirus in contaminated food. Appl Environ Microbiol. 2008;74:4226–4230.

[73] Yan Y, Wang HH, Gao L et al. A one-step multiplex real-time RT-PCR assay for rapid and simultaneous detection of human norovirus genogroup I II and IV. J Virol Methods. 2013;189: 277–282.

[74] Jothikumar N, Cromeans TL, Sobsey MD, Robertson BH. Development andevaluation of a broadly reactive TaqMan assay for rapid detection of hepatitis A virus. Appl Environ Microbiol. 2005;71,3359–3363.

[75] Costafreda MI, Bosch A, Pintó RM. Development, evaluation, and standardization of a real-time TaqMan reverse transcription-PCR assay for quantification of hepatitis A virus in clinical and shellfish samples. Appl Environ Microbiol. 2006;72:3846-55.

[76] Hu Y and ArsoV I. Nested real-time PCR for hepatitis A detection. Lett Appl Microbiol. 2009 ;49:615–619.

[77] Coudray-Meunier C, Fraisse A, Mokhtari C, Martin-Latil S, Roque-Afonso AM, Perelle S. Hepatitis A virus subgenotyping based on RT-qPCR assays. BMC Microbiol. 2014;14:296.

[78] Mohamed N, Belak S, Hedlund KO, Blomberg J. Experience from the development of a diagnostic single tube real-time PCR for human caliciviruses, norovirus genogroups I and II. J Virol Methods. 2006;132:69–76.

[79] Hymas W, Atkinson A, Stevenson J, Hillyard D. Use of modified oligonucleotides to compensate for sequence polymorphisms in the real-time detection of norovirus. J Virol Methods. 2007; 142:10–14.

[80] Abd El Galil KH, El Sokkary MA, Kheira SM et al. Combined immunomagnetic separation-molecular beacon-reverse transcription-PCR assay for detection of hepatitis A virus from environmental samples. Appl Environ Microbiol. 2004;70:4371–4374.

[81] Rosenfield SI, Jaykus LA. A multiplex reverse transcription polymerase chain reaction method for the detection of foodborne viruses. J Food Prot. 1999;10:1210–1214.

[82] Morales-Rayas R, Wolffs PF, Griffiths MW. Simultaneous separation and detection of hepatitis A virus and norovirus in produce. Int J Food Microbiol. 2010;139:48–55.

[83] Fuentes C, Guix S, Perez-Rodriguez FJ et al. Standardized multiplex one-step qRT-PCR for hepatitis A virus, norovirus GI and GII quantification in bivalve mollusks and water. Food Microbiol. 2014;40:55-63.

[84] Farkas T, Singh A, Le Guyader FS et al. Multiplex real-time RT-PCR for the simultaneous detection and quantification of GI, GII and GIV noroviruses. J Virol Methods. 2015;223:109—114.

[85] Chen H and Hu Y. Molecular diagnostic methods for detection and characterization of human noroviruses. 2015 submitted.

[86] Coudray-Meunier C, Fraisse A, Martin-Latil S, Guillier L, Delannoy S, Fach P, Perelle S. A comparative study of digital RT-PCR and RT-qPCR for quantification of Hepatitis A virus and Norovirus in lettuce and water samples. Int J Food Microbiol. 2015;201:17—26.

[87] Houde A, Leblanc D, Poitras E, et al. Comparative evaluation of RT-PCR, nucleic acid sequence-based amplification (NASBA) and real-time RT-PCR for detection of noroviruses in faecal material. J Virol Methods. 2006 ;135:167—172.

[88] Patterson SS, Smith MW, Casper ET et al. A nucleic acid sequence-based amplification assay for real-time detection of norovirus genogroup II. J Appl Microbiol. 2006;101:956—963.

[89] Jean J, Blais B, Darveau A, Fliss I. Detection of hepatitis A virus by the nucleic acid sequence-based amplification technique and comparison with reverse transcription-PCR. Appl Environ Microbiol. 2001;67:5593—5600.

[90] Jean J, Blais B, Darveau A, Fliss I. Simultaneous detection and identification of hepatitis A virus and rotavirus by multiplex nucleic acid sequence-based amplification (NASBA) and microtiter plate hybridization system. J Virol Methods. 2002;105:123—132.

[91] Jean J, D'Souza DH, Jaykus LA. Multiplex nucleic acid sequence-based amplification for simultaneous detection of several enteric viruses in model ready-to-eat foods. Appl Environ Microbiol. 2004;70:6603—6610.

[92] Fukuda S, Takao S, Kuwayama M et al. Rapid detection of norovirus from fecal specimens by real-time reverse transcription-loop-mediated isothermal amplification assay. J Clin Microbiol. 2006;44:1376—1381.

[93] Iturriza-Gómara M, Xerry J, Gallimore CI et al. Evaluation of the Loopamp (loop-mediated isothermal amplification) kit for detecting NorovirusRNA in faecal samples. J Clin Virol. 2008 ;42:389—393.

[94] Yoneyama T, Kiyohara T, Shimasaki N et al. Rapid and real-time detection of hepatitis A virus by reverse transcription loop-mediated isothermal amplification assay. J Virol Methods. 2007;145:162—168.

[95] Jääskeläinen AJ, Maunula L. Applicability of microarray technique for the detection of noro- and astroviruses. J Virol Methods. 2006;136:210—216.

[96] Pagotto F, Corneau N, Mattison K, Bidawid S. Development of a DNA microarray for the simultaneous detection and genotyping of noroviruses. J Food Prot. 2008;71:1434—1441.

[97] Ayodeji M, Kulka M, Jackson SA et al. A microarray based approach for the identification of common foodborne viruses. Open Virol J. 2009;3:7—20.

[98] Chen H, Mammel M, Kulka M, Patel I, Jackson S, Goswami BB. Detection and identification of common food-borne viruses with a tiling microarray. Open Virol J. 2011;5:52—59.

[99] Brinkman NE, Fout GS. Development and evaluation of a generic tag array to detect and genotype noroviruses in water. J Virol Methods. 2009;156:8—18.

[100] Kim JM, Kim SY, Park YB, Kim HJ, Min BS, Cho JC, Yang JM, Cho YH, Ko G. Simultaneous detection of major enteric viruses using a combimatrix microarray. J Microbiol. 2012;50:970—977.

[101] Chen H and Huang S (2012). DNA microarray technology for the detection of food-borne viral pathogens. In: Yan X, Juneja VK, Fratamico PM, Smith JL (eds) Omics, microbial modeling, and technologies in foodborne pathogens. DEStech Publications, Lancaster, pp 579—602.

[102] Chen H, Chen X, Hu Y, Yan H. Reproducibility, fidelity, and discriminant validity of linear RNA amplification for microarray-based identification of major human enteric viruses. Appl Microbiol Biotechnol. 2013;97:4129—4139.

[103] Hu Y, Yan H, Mammel M and Chen H. Sequence-independent amplification coupled with DNA microarray analysis for detection and genotyping of noroviruses. AMB Express. 2015;5:69.

[104] Martínez MA, Soto-Del Río Mde L, Gutiérrez RM et al. DNA microarray for detection of gastrointestinal viruses. J Clin Microbiol. 2015;53:136—145.

[105] Nuanualsuwan S and Cliver DO. Pretreatment to avoid positive RT-PCR results with inactivated viruses. J Virol Methods. 2002;104: 217—225.

[106] Li D, Baert L, Van Coillie E, Uyttendaele M. Critical studies on binding-based RT-PCR detection of infectious Noroviruses. J Virol Methods. 2011;177:153—159.

[107] Dancho BA, Chen H, Kingsley DH. Discrimination between infectious and non-infectious human norovirus using porcine gastric mucin. Int J Food Microbiol. 2012;155:222—226.

[108] Wolf S, Rivera-Aban M, Greening GG. Long-range reverse transcription as a useful tool to assess the genomic integrity of norovirus. Food Environ Virol. 2009;1:129—136.

[109] Parshionikar SU, Cashdollar J, Fout GS. Use of propidium monoazide in reveerse transcriptase PCR to distinguish between infectious and noninfectious enteric viruses in water samples. Appl Envirom Microbiol. 2010;76:4318—4326.

[110] Sánchez G, Elizaquível P, Aznar R. Discrimination of infectious hepatitis A viruses by propidium monoazide real-time RT-PCR. Food Environ Virol. 2012;4:21—25.

[111] Karim MR, Fout GS, Johnson CH, White KM, Parshionikar SU. Propidium monoazide reverse transcriptase PCR and RT-qPCR for detecting infectious enterovirus and norovirus. J Virol Methods. 2015;219:51—61.

[112] Lamhoujeb S, Fliss I, Ngazoa SE, Jean J. Evaluation of the persistence of infectious human noroviruses on food surfaces by using real-time nucleic acid sequence-based amplification. Appl Environ Microbiol. 2008;74:3349—3355.

Chapter 4

Nucleic Acid-based Diagnosis and Epidemiology of Infectious Diseases

Márcia Aparecida Sperança, Rodrigo Buzinaro Suzuki,
Aline Diniz Cabral and Andreia Moreira dos Santos Carmo

Additional information is available at the end of the chapter

http://dx.doi.org/10.5772/61965

Abstract

In this chapter, the immense contribution of nucleic acid discovery to the diagnosis and molecular epidemiology of pathogenic microorganisms and its relevance for veterinary and human health will be discussed. The development of nucleic acid detection, amplification, and sequencing techniques, principally after the introduction of polymerase chain reaction (PCR), allowed the improvement of different strategies to diagnose and to quantify infectious microbiological agents in a variety of organisms and biological samples. Pos-PCR associated techniques such as fragment enzyme restriction and sequence analysis permit the determination of nucleic acid sequence diversity to detect drug resistance, to associate pathogen genetic markers with disease outcome, and to predict temporal and spatial distribution of microorganisms which can be used to prevent and treat infectious diseases efficiently.

The principal methods used in the detection of nucleic acids, the advantages and drawbacks of single- and multiple-copy genes for use in diagnosis by amplification, and the application of pos-PCR techniques in drug resistance identification are discussed in Section 1.1. Section 1.2 discusses the sequencing methods used to recognize genetic variability, the implication of this variability to pathology and virulence, and the importance of genetic variability determination in disease control and vaccines. The contribution of molecular diagnosis and epidemiology for the treatment and prevention of infectious diseases is also considered.

Keywords: Infectious diseases, molecular diagnosis, PCR, real-time PCR, molecular epidemiology, sequencing methods

1. Introduction

1.1. Diagnosis

Several infectious diseases caused by microorganisms, including protozoans (malaria, leishmaniasis, trypanosomiasis, amoebiasis, etc.), bacteria (cholera, gastritis and gastric ulcers, meningitis, tuberculosis, leprosy, etc.), and viruses (dengue fever, yellow fever, influenza, chikungunya fever, ebola, human immunodeficiency syndrome, etc.), are threats to public health. In order to control outbreaks, emergence, and reemergence of these infectious diseases, diagnosis, correct identification, treatment, and notification of pathogenic agents are necessary.

Classical clinical microbiological diagnostics for protozoan and bacteria rely on microscopic examination with different staining methods, culture isolation, morphological and physiological/biochemical characterization. For viruses, the conventional diagnosis is based on culture isolation of cell monolayer, serological assays, and electronic microscopic examination. These standard diagnostic methods are very useful, and culture isolation associated with other analytical procedures to identify microorganisms continues to be the gold standard method since it enables drug sensitivity tests. However, these diagnostic techniques are unsuitable for several microorganisms presenting fastidious growth characteristics, low morphological and physiological specificity, and requirement of specific biosafety infrastructure. The same is applied to viruses that, even after culture isolation, can only be visualized by electronic microscopy, which is expensive and needs specialized personnel to maneuver. Thus, nucleic acid detection by hybridization and amplification technologies opened a new and innovative period for microbial diagnosis. After the first report on the application of polymerase chain reaction (PCR) in clinical diagnosis of the human immunodeficiency virus (HIV) [1], several other infectious organisms were detected by the same technique and its variations (Section 1.1.1).

In all molecular detection techniques, the gene target is the main device, and its choice depends on the infectious agent and the host genomic and epidemiological characteristics. For a specific diagnosis, the gene of choice has to be specific to the infectious agent and should not cross-hybridize with the host genome and other organisms living in the same microhabitat. A sensitive diagnosis depends on the amount of gene target copies in the biological sample and to the physico-chemical characteristics of the constituents implicated in the detection and amplification of gene target. A discussion about the use of single- and multiple-copy genes for specific gene target amplification will be presented in Section 1.1.2.

The detection of drug resistance is dependent on sensitivity tests performed on the isolated microorganism, which is time consuming; however, for several uncultivable pathogenic agents, it is not feasible. The investigation of nucleotide mutations associated to drug resistance allows the development of gene target amplification and post-amplification analytical techniques, such as enzyme restriction analysis and sequencing, to be used directly on the biological infected sample, thereby enabling fast detection of drug resistance and consequently an efficient treatment. The same strategy can be used to identify

http://dx.doi.org/10.5772/61965

organisms from closed biological groups, with identical morphological characteristics on microscopic examination and with different genetic features. Examples of nucleic acid drug resistance detection techniques used in microbiological clinical laboratories will be presented in Section 1.1.3.

1.1.1. Nucleic acid detection methods

Originally, the nucleic acids were detected mainly by gene cloning strategies and hybridization procedures [2], which are laborious and time consuming, being restricted to scientific investigation. The use of nucleic acid detection for the diagnosis of genetic and infectious diseases in clinical laboratories was possible after the advent of PCR [3], a technique based on amplification of nucleic acids by means of thermostable polymerase enzymes and a thermocycler. By this method, typically, DNA duplex templates are melted at high temperatures, and two oligonucleotides complementary to the flanking gene target sequence are specifically annealed in a strict temperature dependent on the primer sequence and length. Variations of the technique includes the utilization of a set of oligonucleotides in order to identify different organisms or variants in a single reaction on a biological sample [4-7]. Also, to increase sensitivity and specificity, a nested and hemi/semi-nested PCR can be performed using an initial PCR product as template [8, 9, 6, 10].

Besides PCR, the most largely used nucleic acid amplification device, isothermal amplification techniques based on enzymes required during the cellular process of DNA/RNA synthesis were also developed and are accessible for diagnosis and scientific investigation as transcription-mediated amplification (TMA), nucleic acid sequence-based amplification (NASBA), signal-mediated amplification of RNA technology (SMART), strand displacement amplification (SDA), rolling circle amplification (RCA), loop-mediated isothermal amplification of DNA (LAMP), isothermal multiple displacement amplification (IMDA), helicase-dependent amplification (HDA), single primer isothermal amplification (SPIA), and circular helicase-dependent amplification (cHDA). The description of these methods is out of the scope of this chapter, and several revision articles can be consulted for more information [11-13].

Subsequent to amplification, PCR products are traditionally visualized by an electrophoresis in agarose or acrylamide gels following staining with fluorescent dyes and exposition to UV light. By this method, the specificity of the amplified nucleic acid fragment is determined by size, directly or after digestion with restriction enzymes in order to get more accurate information about the product obtained. The specificity of PCR products can also be determined by sequencing (Section 1.2.2) and hybridization with chemically, fluorescently, or radioactively labeled specific probes as exemplified by the means of human papillomavirus (HPV) genotyping [14, 15].

A novel improvement in nucleic acid amplification was achieved after the development of PCR in the presence of fluorescent dyes, enabling the detection of products by amplification cycle and, at real time, the real-time PCR [16]. Thermocycles developed for the real-time PCR are associated to a fluorescence detection system and software to facilitate interpretation of data at real time. Although conventional PCR allows the amplification of DNA fragments as long as 20 Kb, the size of DNA fragment obtained by the real-time PCR is not longer than 150 bp which cannot be used for pos-PCR analysis. The specificity of a real-time PCR product is determined by the use of Carl Wittwer's melting curve analysis [17] or by using dual fluorescently labeled probes [18]. These improvements allowed the use of real-time PCR not only to detect genotype [19, 20] but also to quantify the amplification product, to determine gene copy numbers of pathogenic microorganisms and the expression of genes associated to infection reactivation, virulence, genetic modification, etc. [21-24].

1.1.2. Advantages and drawbacks of single and multiple copy genes

The specificity and sensitivity of nucleic acid amplification techniques depend on the target gene selection and the design of oligonucleotides. Once the gene target is elected, there are several free available bioinformatic tools to check for the better sequence to be used in PCR and real-time PCR reactions, mainly to avoid the formation of oligonucleotide self-hairpin structure and primer–dimer arrangements. Thus, the principal challenge to obtain an accurate molecular diagnosis based on nucleic acid amplification is the selection of the target genomic sequence.

The biological features of the microorganism to be diagnosed are of critical importance to direct the choice of an accurate gene target, including the life cycle, specific metabolic pathways, genomic organization, and evolution. Knowledge on life cycle is principally important when the purpose is the detection of stage-specific forms of a microorganism, as gametocytes of *Plasmodium* in human blood [25]. In this case, the real-time PCR can detect the expression of gametocyte-specific genes, and patients at risk to transmit malaria can be identified.

Particular metabolic pathways of a microorganism are usually associated to specific genes which can be used as markers for the recognition of infectious agents [26-28]. Commonly, these genes are in single copy in the genome and are highly expressed and evolutionarily conserved, being excellent targets for real-time PCR diagnosis [29], as it is a highly sensitive nucleic acid detection method.

Genomic organization and evolution are of principal interest to select gene targets that are not in strong structured regions of the genome [30] and whose sequence is beneath strong selective pressure, being highly conserved. Usually, multiple-copy genes as ribosomal and transporter RNA coding sequences are used as template for diagnostic methods in order to obtain high sensitivity with specificity. In this case, it is important to consider that RNA coding sequences possess a high capability to form hairpins and are in a structured region of the genome, and thus, even in high copy number, they can present a low sensitivity in PCR reactions. Also, as ribosomal and transporter RNA coding sequences are very conserved among all groups of organisms, decreasing sensitivity of the diagnostic test can occur depending on the selected gene region. Thus, it is very important to perform a comparative analysis of the diagnostically selected gene among phylogenetically closed groups of organisms and the host. Mitochondrial genes are also largely used as gene targets due to their high copy number in eukaryotes; however, their similarity with bacterial genome also has to be observed in order to obtain specificity [31].

http://dx.doi.org/10.5772/61965

1.1.3. Identification of drug resistance

After amplification of a target gene by PCR, several molecular strategies can be used to detect nucleotide mutations, such as sequencing, restriction enzyme analysis, hybridization, etc. This approach is useful to detect mutations associated to drug resistance directly on biological samples without the requirement of culture isolation. Some antibiotics act in bacterial ribosome, and the investigation of point mutations in ribosomal RNA coding sequences of cultivable bacteria can be extended by Homology to fastidious growing pathogenic species. One example is the fastidious cultivable bacteria *Helicobacter pylori,* the etiological agent of gastritis and peptic ulcer diseases. There are several real-time PCR and pos-PCR methods for clinical applications in order to determine resistance against the most important antibiotics used for *H. pylori* treatment, including clarithromycin [32, 33] and tetracycline [34, 35].

The same kind of methodology is used to investigate genotypes of microorganisms associated to pathogenicity, virulence, drug resistance, etc. As an example, genotype identification of HIV [36, 37] and hepatitis C virus (HCV) [38] resistant to treatment with different inhibitors of viral protease, directly on clinical samples, allow correct treatment of the patients. Another example is the differentiation of the protozoa *Entamoeba histolytica* (pathogenic) from *E. dispar* (non-pathogenic) and *E. moshkovskii* (non-pathogenic) in the intestinal tract of human, which can only be performed genetically [39]. The correct identification of *E. histolytica* avoids unnecessary treatment which, besides being expensive and capable of causing side effects to the patient, permits selection of drug-resistant species.

1.2. Epidemiology

Nucleic acid approaches have improved epidemiology, a science that deals with etiology, distribution, natural history, and control of diseases in humans [40]. In this section, some aspects of the molecular epidemiology of infectious diseases will be discussed, including the determination of etiological agents of diseases, association of genetic markers to transmission, treatment efficiency, clinical outcome patterns, applicability of genetic diversity knowledge in vaccine development, and the control/prevention of transmission of infectious agents.

Several genotyping techniques are mentioned in Section 1.1, and the same approaches can be used in order to determine the genetic variability. However, the development of enzymatic nucleic acid sequence determination with the use of chain-terminating inhibitors by Sanger [41] and its rapid automation allowed a prompt availability of gene sequences of several infectious agents and their hosts. More recently, next-generation sequencing methods associated to powerful bioinformatic tools have made complete sequences of small genomes of pathogenic virus and bacteria accessible during the occurrence of an epidemic or outbreak. These techniques will be briefly presented in Section 1.2.1.

Epidemiological studies on the association of pathogens and host genetic variability to disease susceptibility and pathogenicity/virulence can help to prevent and treat several infectious diseases (Section 1.2.2.). Moreover, the determination of genetic variability of infectious agents, antigens, and host population are useful for vaccine development (Section 1.2.3).

1.2.1. Classic genotyping and next-generation sequencing methods

Different from the diagnostic methods, where molecular markers are desired to be conserved among all individuals of a species, for population genetic epidemiological studies, the most important marker characteristics are individual high variability and neutral evolution. Nucleic acid-based conventional genotyping of microorganisms can be carried out by hybridization and amplification approaches as already described in Section 1.1. Moreover, the most classical molecular approaches used for epidemiological studies include restriction fragment length polymorphism (RFLP), macrorestriction analysis [42], and alternative PCR and pos-PCR techniques such as rapid amplification of polymorphic DNA (RAPD), analysis of variable number of tandem repeats (VNTR) [43], variation in repeated short motifs (microsatellites), multilocus microsatellite typing (MLMT) [44, 45], single-strand conformation polymorphism (SSCP) [46], etc. These techniques are valuable for epidemiological studies of bacteria and protozoans, considering their genomic organization, structure, and large size. The detection of single-nucleotide polymorphism (SNP) associated to specific populational characteristics can be easily identified by sequencing. Also, in majority of viruses, which generally present a small genome and lack of repetitive tandem organized sequences, complete genomic sequencing is the genotyping method of choice for epidemiologic studies.

Genomic sequencing became possible for many biological applications after the development of first-generation sequencing methods, comprising Maxam–Gilbert's base-specific chemical cleavage [47] and Sanger's enzymatic reaction with chain terminator nucleotides, which is widely used [41]. The improvement of Sanger sequencing method with the use of thermostable proofreading DNA polymerases, fluorescent chain terminator nucleotides, and laser-based equipment associated to capillary electrophoresis enabled its rapid automation and commercial availability. Therefore, Sanger's method allowed a revolution in genome sequence projects of pathogenic microorganisms and other organisms, with high accuracy and a relatively low rate [48-53]. The search for a more throughput sequencing technique to obtain large complete genomic sequences, more rapidly and without purification or cloning of the nucleic acid of a specific organism, leads to the development of second- and third-generation sequencing methods.

The second-generation method was commercialized by Roche laboratories, the 454 Life Science equipment, which was based on pyrophosphate detection during DNA synthesis performed on an array composed of wells in a fiber-optic slide [54]. By this method, each well is filled with one bead containing a single DNA molecule polymer produced by genomic sharing, and primed at 3' and 5' ends with short sequence adaptors. DNA polymer attached to a bead is immersed in an emulsion to be amplified by PCR in order to obtain a high number of molecules to generate enough luminescence signals to be captured by a charge-coupled device (CCD) imager coupled to the fiber-optic array. Before its attachment to the well, each amplified DNA polymer is denatured with specific enzymes. Each sequence cycle consists of pyrophosphate detection in each fiber-optic well, released after the addition of one of the four DNA bases and captured by a luminescent reaction system and read at real time. Each well of the array produces a collection of short sequences of approximately 40 bp which are assembled by software with the use of a prototype Sanger-generated genomic sequence available in several public databanks. Roche laboratories discontinued the 454 Life Science which was replaced by the improved FLX titanium XLR and XL+ capable of individual length read of 450 and 700 bp, respectively. Other second-generation systems, characterized by the production of short sequence arrays, are based on DNA synthesis with fluorescent terminator reversible nucleotides and exonucleases. A CCD imager collects the enzymatically cleaved incorporated specific terminator fluorescent nucleotide, added by cycles of sequencing on primed DNA fragments, produced by sharing and attached on a solid platform, where it is also amplified by PCR [55]. These technologies present certain limitations such as a relative high cost, low throughput, and low sequence accuracy for large genomes, although being valuable for virus genomes [56].

Third-generation sequencing methodologies are in development, based on a technology called single molecule real time (SMRT), which use biological nanopores, without the necessity to amplify the target template [57]. All these next-generation sequencing techniques have been improving, and certainly more accessible and accurate devices will be available in near future.

1.2.2. Genetic variability and its impact on pathology and virulence

Genetic variability determination in microorganisms, performed by different methodologies as exemplified in Section 1.2.1, allows the identification of infectious disease distribution in time and place, as well as the investigation of transmission patterns and clinical manifestation and progression in affected populations. Knowledge of these factors is used for infectious disease intervention and prevention strategies, as for example in dengue fever, caused by four distinct serotypes of dengue virus (DENV1-4), belonging to the genus Flavivirus, family Flaviviridae.

The clinical manifestation and progression of dengue patients vary from absence of symptoms to a severe disease, mostly characterized by plasma leakage with or without hemorrhagic signs [58]. According to several studies, symptoms and severity of dengue diseases are associated to viral load, host immunity conditions, and the occurrence of antibody-dependent enhancement by heterologous DENV antibodies which are related to the number of previous serotypes infections. DENV serotypes are related to genetic variation in the major antigenic envelope and pre-membrane structural proteins. Thus, the characterization of DENV genetic variation and evolution in these specific regions is of crucial importance to understand the disease progress in an affected population.

Molecular epidemiological investigations, with sequencing of DENV 240 nucleotides of the envelop and nonstructural 1 coding regions, revealed distinct genomic groups of DENV serotypes 1 and 2, according to the geographical and temporal virus population circulation [59], which was confirmed by complete genome sequencing [60]. The same strategy was used to genetically characterize groups of DENV-3 and DENV-4, with distinct evolutionary constraints which interfere in transmission and disease manifestation [61, 62]. Knowledge of these characteristics has implications not only for control and disease management strategies but also for vaccine development.

Microsatellite genomic sequences are useful for the treatment and control of infectious agents by the identification of individual and group of clonally disseminated microorganisms. This kind of study enables the identification of relapses, drug-resistant pathogens, and clinically distinct variants, as investigated in protozoan parasites such as Leishmania [45, 63] and Plasmodium [64, 65]. The geographical dissemination of microorganisms can also be characterized by microsatellites studies [66, 67] helping in the characterization of transmission patterns which can be used for control strategies.

Domestic and sylvatic animal sources of pathogenic microorganisms infecting human, with or without clinical outcomes according to genetic variants, can also be investigated by genotyping. Toxoplasmosis, an example of a zoonosis, is caused by *Toxoplasma gondii*, a protozoan parasite with high genetic variability, from the apicomplexa group. Attempts to identify genetic variants associated to specific animal hosts or clinical outcome fail; however, it is possible to identify the source of parasite infection by genotyping with the use of multilocus PCR and RFLP [68]. Using this methodology, a study on German patients revealed that ocular, cerebral, and systemic toxoplasmosis presented predominance of *T. gondii* from lineage type II, the same genotype found in parasite oocysts isolated from cats of the same geographic region [69]. The prevalence of *T. gondii* lineages presents specific regional distribution, being highly variable in South America [70], where the parasites are isolated from domestic and wild animals, including chiropterans [71]. Based on these studies, it is possible to recognize several potential parasite intermediate hosts and map the risk areas for toxoplasmosis transmission.

Chaga's disease caused by *Trypanosoma cruzi*, a protozoan parasite from the kinetoplastida group, is also an excellent example of eco-epidemiologic study using genotyping. Parasites are transmitted naturally to humans by the exposure of abraded skin to feces of triatomine bugs, and they circulate in the wild in different sylvatic animals and triatomine species. There are six principal genetic variants or discrete typing units of *T. cruzi* determined by multilocus PCR, RFLP, and sequencing analysis, associated to geographical distribution and transmission cycles [72-74]. The association of the outcome of Chaga's disease and specific parasite genotypes is still controversial; however, comparative investigation of *T. cruzi* genotypes in human infections, sylvatic bugs, and animal reservoirs, in rural areas near forest, which is common in South America, can reveal potential vectors and animal sources of infection.

1.2.3. Disease control and vaccines

The most effective means to control an infectious disease is a vaccine, which is still not available for malaria, trypanosomiasis (Chaga's disease, sleeping sickness, leishmaniasis), and most of arboviruses (dengue fever, chikungunya fever, etc.). The genetic variability of infectious agents, a rapid genomic evolution in case of viruses, and several escape mechanisms from host immune system selected along parasite–host interactions constitutes intricate drawbacks for vaccine development. The problem of genetic prompt variability of virus genome can be circumvented by systematic sequencing during epidemic periods such as the one successfully made for influenza vaccine recommendation each year.

Influenza vaccine is composed of attenuated viruses produced in eggs, every year, following the virus strain recommendation by World Health Organization, after monitoring genetic variation in the gene encoding the surface protein hemagglutinin due to its antigenic properties (http://www.who.int/influenza/vaccines/virus/recommendations/201502_recommendation.pdf?ua=1). Molecular epidemiological studies performed in different geographical regions are used in order to obtain the recommended vaccine strains. Also, there are geographical and populational differences in viral antigenic changes which should be taken into consideration as described in China, where dominant antigenic influenza clusters change more frequently than in the USA/Europe [75].

Dengue virus consists of a particular challenging agent for vaccine development since its genetic variation is directly related to the severity of disease progress, due to the antibody-dependent enhancement by heterologous previous DENV infection. Thus, geographical regional circulation knowledge of DENV virus is essential to the success of a vaccine [61, 76]. Considering the high speed in data acquisition, second- and third-generation sequencing of DENV circulating during interepidemic period could help to predict serotypes and its genetic variants in next outbreak, contributing to vaccine development and testing.

2. Final Considerations

Economic and social impact of infectious diseases should be considered worldwide, principally in developing countries, including necessity of medical care, vector control, people morbidity with loss of working hours, and reduced tourism in affected areas. The increase in population, high people mobility, and the lack of an effective vaccine against the most disseminated infectious diseases, such as dengue fever, malaria, tripanosomiasis, etc., make these human threats an important public health concern. Nucleic acid technologies have been contributing to the knowledge of infectious disease transmission, distribution, clinical manifestations, and progression, helping in its control and management. Several examples are discussed in this chapter, and hopefully, it will stimulate readers to investigate these microbiological intricacies and magnificent parasite–host interactions, more profoundly.

Author details

Márcia Aparecida Sperança[1*], Rodrigo Buzinaro Suzuki[1,2], Aline Diniz Cabral[1] and Andreia Moreira dos Santos Carmo[1,3]

*Address all correspondence to: marcia.speranca@ufabc.edu.br

1 Universidade Federal do ABC, Center for Natural and Human Sciences, São Bernardo do Campo, São Paulo, Brazil

2 Department of Genotyping, Hemocenter, Marilia Medical School, Marilia, São Paulo, Brazil

3 Secretaria do Estado da Saúde do Estado de São Paulo, Instituto Adolfo Lutz, Centro de Laboratório Regional VIII, Santo André, São Paulo, Brazil

References

[1] Kwok S, Mack DH, Mullis KB, Poiesz B, Ehrlich G, Blair D et al. Identification of human immunodeficiency virus sequences by using in vitro enzymatic amplification and oligomer cleavage detection. Journal of virology. 1987;61(5):1690-4.

[2] Tsongalis GJ, Silverman LM. Molecular diagnostics: a historical perspective. Clinica chimica acta; international journal of clinical chemistry. 2006;369(2):188-92. doi: 10.1016/j.cca.2006.02.044.

[3] Saiki RK, Scharf S, Faloona F, Mullis KB, Horn GT, Erlich HA et al. Enzymatic amplification of beta-globin genomic sequences and restriction site analysis for diagnosis of sickle cell anemia. Science. 1985;230(4732):1350-4.

[4] Mishra B, Sharma M, Pujhari SK, Appannanavar SB, Ratho RK. Clinical applicability of single-tube multiplex reverse-transcriptase PCR in dengue virus diagnosis and serotyping. Journal of clinical laboratory analysis. 2011;25(2):76-8. doi:10.1002/jcla.20434.

[5] Bronzoni RV, Moreli ML, Cruz AC, Figueiredo LT. Multiplex nested PCR for Brazilian Alphavirus diagnosis. Trans R Soc Trop Med Hyg. 2004;98(8):456-61. doi:10.1016/j.trstmh.2003.09.002.

[6] Kourenti C, Karanis P. Evaluation and applicability of a purification method coupled with nested PCR for the detection of Toxoplasma oocysts in water. Letters in applied microbiology. 2006;43(5):475-81. doi:10.1111/j.1472-765X.2006.02008.x.

[7] Kim S, Lee DS, Suzuki H, Watarai M. Detection of Brucella canis and Leptospira interrogans in canine semen by multiplex nested PCR. The Journal of veterinary medical science / the Japanese Society of Veterinary Science. 2006;68(6):615-8.

[8] Gomes AL, Silva AM, Cordeiro MT, Guimaraes GF, Marques ET, Jr., Abath FG. Single-tube nested PCR using immobilized internal primers for the identification of dengue virus serotypes. Journal of virological methods. 2007;145(1):76-9. doi:10.1016/j.jviromet.2007.05.003.

[9] Tuksinvaracharn R, Tanayapong P, Pongrattanaman S, Hansasuta P, Bhattarakosol P, Siriyasatien P. Prevalence of dengue virus in Aedes mosquitoes during dry season by semi-nested reverse transcriptase-polymerase chain reaction (semi-nested RT-PCR). Journal of the Medical Association of Thailand = Chotmaihet thangphaet. 2004;87 Suppl 2:S129-33.

[10] Hurtado A, Aduriz G, Moreno B, Barandika J, Garcia-Perez AL. Single tube nested PCR for the detection of Toxoplasma gondii in fetal tissues from naturally aborted ewes. Veterinary parasitology. 2001;102(1-2):17-27.

[11] Craw P, Balachandran W. Isothermal nucleic acid amplification technologies for point-of-care diagnostics: a critical review. Lab on a chip. 2012;12(14):2469-86. doi: 10.1039/c2lc40100b.

[12] Asiello PJ, Baeumner AJ. Miniaturized isothermal nucleic acid amplification, a review. Lab on a chip. 2011;11(8):1420-30. doi:10.1039/c0lc00666a.

[13] Gill P, Ghaemi A. Nucleic acid isothermal amplification technologies: a review. Nucleosides, nucleotides & nucleic acids. 2008;27(3):224-43. doi: 10.1080/15257770701845204.

[14] Poonnaniti A, Bhattarakosol P. Improvement of PCR detection of HPV-DNA using enhanced chemiluminescence system and dot hybridization. Journal of the Medical Association of Thailand = Chotmaihet thangphaet. 1996;79 Suppl 1:S96-103.

[15] Moller B, Lindeberg H. The presence of HPV types 6/11 and 16 in a giant vulvar carcinoma, arising in pre-existing condylomas, demonstrated by PCR and DNA in-situ hybridization. European journal of gynaecological oncology. 1996;17(6):497-500.

[16] Gingeras TR, Higuchi R, Kricka LJ, Lo YM, Wittwer CT. Fifty years of molecular (DNA/RNA) diagnostics. Clinical chemistry. 2005;51(3):661-71. doi:10.1373/clinchem.2004.045336.

[17] Pryor RJ, Wittwer CT. Real-time polymerase chain reaction and melting curve analysis. Methods Mol Biol. 2006;336:19-32. doi:10.1385/1-59745-074-X:19.

[18] Holland PM, Abramson RD, Watson R, Gelfand DH. Detection of specific polymerase chain reaction product by utilizing the 5'----3' exonuclease activity of Thermus aquaticus DNA polymerase. Proceedings of the National Academy of Sciences of the United States of America. 1991;88(16):7276-80.

[19] Athar MA, Xu Y, Xie X, Xu Z, Ahmad V, Hayder Z et al. Rapid detection of HCV genotyping 1a, 1b, 2a, 3a, 3b and 6a in a single reaction using two-melting temperature codes by a real-time PCR-based assay. Journal of virological methods. 2015;222:85-90. doi:10.1016/j.jviromet.2015.05.013.

[20] Zhou L, Gong R, Lu X, Zhang Y, Tang J. Development of a multiplex real-time PCR assay for the detection of Treponema pallidum, HCV, HIV-1 and HBV. Japanese journal of infectious diseases. 2015. doi:10.7883/yoken.JJID.2014.416.

[21] Gaydos CA. Review of use of a new rapid real-time PCR, the Cepheid GeneXpert(R) (Xpert) CT/NG assay, for Chlamydia trachomatis and Neisseria gonorrhoeae: results for patients while in a clinical setting. Expert review of molecular diagnostics. 2014;14(2):135-7. doi:10.1586/14737159.2014.871495.

[22] Ngo-Giang-Huong N, Khamduang W, Leurent B, Collins I, Nantasen I, Leechanachai P et al. Early HIV-1 diagnosis using in-house real-time PCR amplification on dried blood spots for infants in remote and resource-limited settings. J Acquir Immune Defic Syndr. 2008;49(5):465-71. doi:10.1097/QAI.0b013e31818e2531.

[23] Albuquerque YM, Lima AL, Lins AK, Magalhaes M, Magalhaes V. Quantitative real-time PCR (q-PCR) for sputum smear diagnosis of pulmonary tuberculosis among people with HIV/AIDS. Rev Inst Med Trop Sao Paulo. 2014;56(2):139-42. doi:10.1590/S0036-46652014000200009.

[24] Strassl R, Rutter K, Stattermayer AF, Beinhardt S, Kammer M, Hofer H et al. Real-Time PCR Assays for the Quantification of HCV RNA: Concordance, Discrepancies and Implications for Response Guided Therapy. PloS one. 2015;10(8):e0135963. doi: 10.1371/journal.pone.0135963.

[25] Jones S, Sutherland CJ, Hermsen C, Arens T, Teelen K, Hallett R et al. Filter paper collection of Plasmodium falciparum mRNA for detecting low-density gametocytes. Malaria journal. 2012;11:266. doi:10.1186/1475-2875-11-266.

[26] Goldberg DE, Slater AF, Beavis R, Chait B, Cerami A, Henderson GB. Hemoglobin degradation in the human malaria pathogen Plasmodium falciparum: a catabolic pathway initiated by a specific aspartic protease. The Journal of experimental medicine. 1991;173(4):961-9.

[27] Rosowski EE, Lu D, Julien L, Rodda L, Gaiser RA, Jensen KD et al. Strain-specific activation of the NF-kappaB pathway by GRA15, a novel Toxoplasma gondii dense granule protein. The Journal of experimental medicine. 2011;208(1):195-212. doi: 10.1084/jem.20100717.

[28] Fiocca R, Necchi V, Sommi P, Ricci V, Telford J, Cover TL et al. Release of Helicobacter pylori vacuolating cytotoxin by both a specific secretion pathway and budding of outer membrane vesicles. Uptake of released toxin and vesicles by gastric epithelium. The Journal of pathology. 1999;188(2):220-6. doi:10.1002/(SICI)1096-9896(199906)188:2<220::AID-PATH307>3.0.CO;2-C.

[29] Gayet-Ageron A, Combescure C, Lautenschlager S, Ninet B, Perneger TV. Comparison of the diagnostic accuracy of polymerase chain reaction targeting the 47kDa protein membrane gene of Treponema pallidum or the DNA polymerase I gene: a systematic review and meta-analysis. Journal of clinical microbiology. 2015. doi: 10.1128/JCM.01619-15.

[30] Hatfield GW, Benham CJ. DNA topology-mediated control of global gene expression in Escherichia coli. Annual review of genetics. 2002;36:175-203. doi:10.1146/annurev.genet.36.032902.111815.

[31] Opota O, Jaton K, Branley J, Vanrompay D, Erard V, Borel N et al. Improving the molecular diagnosis of Chlamydia psittaci and Chlamydia abortus infection with a species-specific duplex real-time PCR. Journal of medical microbiology. 2015. doi: 10.1099/jmm.0.000139.

[32] Tankovic J, Chaumette-Planckaert MT, Deforges L, Launay N, Le Glaunec JM, Soussy CJ et al. Routine use of real-time PCR for detection of Helicobacter pylori and of

clarithromycin resistance mutations. Gastroenterologie clinique et biologique. 2007;31(10):792-5.

[33] Suzuki RB, Lopes RA, da Camara Lopes GA, Hung Ho T, Speranca MA. Low Helicobacter pylori primary resistance to clarithromycin in gastric biopsy specimens from dyspeptic patients of a city in the interior of Sao Paulo, Brazil. BMC gastroenterology. 2013;13:164. doi:10.1186/1471-230X-13-164.

[34] Ribeiro ML, Gerrits MM, Benvengo YH, Berning M, Godoy AP, Kuipers EJ et al. Detection of high-level tetracycline resistance in clinical isolates of Helicobacter pylori using PCR-RFLP. FEMS immunology and medical microbiology. 2004;40(1):57-61.

[35] Dadashzadeh K, Milani M, Rahmati M, Akbarzadeh A. Real-time PCR detection of 16S rRNA novel mutations associated with Helicobacter pylori tetracycline resistance in Iran. Asian Pacific journal of cancer prevention : APJCP. 2014;15(20):8883-6.

[36] Steegen K, Demecheleer E, De Cabooter N, Nges D, Temmerman M, Ndumbe P et al. A sensitive in-house RT-PCR genotyping system for combined detection of plasma HIV-1 and assessment of drug resistance. Journal of virological methods. 2006;133(2): 137-45. doi:10.1016/j.jviromet.2005.11.004.

[37] Schmit JC, Ruiz L, Stuyver L, Van Laethem K, Vanderlinden I, Puig T et al. Comparison of the LiPA HIV-1 RT test, selective PCR and direct solid phase sequencing for the detection of HIV-1 drug resistance mutations. Journal of virological methods. 1998;73(1):77-82.

[38] Silva T, Cortes Martins H, Coutinho R, Leitao E, Silva R, Padua E. Molecular characterization of hepatitis C virus for determination of subtypes and detection of resistance mutations to protease inhibitors in a group of intravenous drug users co-infected with HIV. Journal of medical virology. 2015;87(9):1549-57. doi:10.1002/jmv.24213.

[39] Zebardast N, Haghighi A, Yeganeh F, Seyyed Tabaei SJ, Gharavi MJ, Fallahi S et al. Application of Multiplex PCR for Detection and Differentiation of Entamoeba histolytica, Entamoeba dispar and Entamoeba moshkovskii. Iranian journal of parasitology. 2014;9(4):466-73.

[40] Lilienfeld DE. Definitions of epidemiology. American journal of epidemiology. 1978;107(2):87-90.

[41] Sanger F, Nicklen S, Coulson AR. DNA sequencing with chain-terminating inhibitors. Proceedings of the National Academy of Sciences of the United States of America. 1977;74(12):5463-7.

[42] Tiong V, Thong KL, Yusof MY, Hanifah YA, Sam JI, Hassan H. Macrorestriction analysis and antimicrobial susceptibility profiling of Salmonella enterica at a University Teaching Hospital, Kuala Lumpur. Japanese journal of infectious diseases. 2010;63(5): 317-22.

[43] Ramazanzadeh R, McNerney R. Variable Number Of Tandem Repeats (VNTR) and its application in bacterial epidemiology. Pakistan journal of biological sciences : PJBS. 2007;10(16):2612-21.

[44] Montoya L, Gallego M, Gavignet B, Piarroux R, Rioux JA, Portus M et al. Application of microsatellite genotyping to the study of a restricted Leishmania infantum focus: different genotype compositions in isolates from dogs and sand flies. The American journal of tropical medicine and hygiene. 2007;76(5):888-95.

[45] Aluru S, Hide M, Michel G, Banuls AL, Marty P, Pomares C. Multilocus microsatellite typing of Leishmania and clinical applications: a review. Parasite. 2015;22:16. doi: 10.1051/parasite/2015016.

[46] Glavac D, Dean M. Optimization of the single-strand conformation polymorphism (SSCP) technique for detection of point mutations. Human mutation. 1993;2(5): 404-14. doi:10.1002/humu.1380020513.

[47] Maxam AM, Gilbert W. A new method for sequencing DNA. Proceedings of the National Academy of Sciences of the United States of America. 1977;74(2):560-4.

[48] Sasaki Y, Ishikawa J, Yamashita A, Oshima K, Kenri T, Furuya K et al. The complete genomic sequence of Mycoplasma penetrans, an intracellular bacterial pathogen in humans. Nucleic Acids Res. 2002;30(23):5293-300.

[49] Tettelin H, Masignani V, Cieslewicz MJ, Eisen JA, Peterson S, Wessels MR et al. Complete genome sequence and comparative genomic analysis of an emerging human pathogen, serotype V Streptococcus agalactiae. Proceedings of the National Academy of Sciences of the United States of America. 2002;99(19):12391-6. doi:10.1073/pnas.182380799.

[50] Gardner MJ, Tettelin H, Carucci DJ, Cummings LM, Smith HO, Fraser CM et al. The malaria genome sequencing project: complete sequence of Plasmodium falciparum chromosome 2. Parassitologia. 1999;41(1-3):69-75.

[51] Ravel C, Dubessay P, Bastien P, Blackwell JM, Ivens AC. The complete chromosomal organization of the reference strain of the Leishmania Genome Project, L. major ;Friedlin'. Parasitol Today. 1998;14(8):301-3.

[52] Tomb JF, White O, Kerlavage AR, Clayton RA, Sutton GG, Fleischmann RD et al. The complete genome sequence of the gastric pathogen Helicobacter pylori. Nature. 1997;388(6642):539-47. doi:10.1038/41483.

[53] Berriman M, Ghedin E, Hertz-Fowler C, Blandin G, Renauld H, Bartholomeu DC et al. The genome of the African trypanosome Trypanosoma brucei. Science. 2005;309(5733):416-22. doi:10.1126/science.1112642.

[54] Ronaghi M, Uhlen M, Nyren P. A sequencing method based on real-time pyrophosphate. Science. 1998;281(5375):363, 5.

[55] Shendure J, Ji H. Next-generation DNA sequencing. Nature biotechnology. 2008;26(10):1135-45. doi:10.1038/nbt1486.

[56] Datta S, Budhauliya R, Das B, Chatterjee S, Vanlalhmuaka, Veer V. Next-generation sequencing in clinical virology: Discovery of new viruses. World journal of virology. 2015;4(3):265-76. doi:10.5501/wjv.v4.i3.265.

[57] Maitra RD, Kim J, Dunbar WB. Recent advances in nanopore sequencing. Electrophoresis. 2012;33(23):3418-28. doi:10.1002/elps.201200272.

[58] . Dengue: Guidelines for Diagnosis, Treatment, Prevention and Control: New Edition. WHO Guidelines Approved by the Guidelines Review Committee. Geneva2009.

[59] Rico-Hesse R. Molecular evolution and distribution of dengue viruses type 1 and 2 in nature. Virology. 1990;174(2):479-93.

[60] Lewis JA, Chang GJ, Lanciotti RS, Kinney RM, Mayer LW, Trent DW. Phylogenetic relationships of dengue-2 viruses. Virology. 1993;197(1):216-24. doi:10.1006/viro.1993.1582.

[61] Lanciotti RS, Lewis JG, Gubler DJ, Trent DW. Molecular evolution and epidemiology of dengue-3 viruses. J Gen Virol. 1994;75 (Pt 1):65-75.

[62] Chen R, Vasilakis N. Dengue--quo tu et quo vadis? Viruses. 2011;3(9):1562-608. doi:10.3390/v3091562.

[63] Krayter L, Bumb RA, Azmi K, Wuttke J, Malik MD, Schnur LF et al. Multilocus microsatellite typing reveals a genetic relationship but, also, genetic differences between Indian strains of Leishmania tropica causing cutaneous leishmaniasis and those causing visceral leishmaniasis. Parasites & vectors. 2014;7:123. doi:10.1186/1756-3305-7-123.

[64] Mallick PK, Sutton PL, Singh R, Singh OP, Dash AP, Singh AK et al. Microsatellite analysis of chloroquine resistance associated alleles and neutral loci reveal genetic structure of Indian Plasmodium falciparum. Infection, genetics and evolution : journal of molecular epidemiology and evolutionary genetics in infectious diseases. 2013;19:164-75. doi:10.1016/j.meegid.2013.07.009.

[65] Thanapongpichat S, McGready R, Luxemburger C, Day NP, White NJ, Nosten F et al. Microsatellite genotyping of Plasmodium vivax infections and their relapses in pregnant and non-pregnant patients on the Thai-Myanmar border. Malaria journal. 2013;12:275. doi:10.1186/1475-2875-12-275.

[66] Gouzelou E, Haralambous C, Antoniou M, Christodoulou V, Martinkovic F, Zivicnjak T et al. Genetic diversity and structure in Leishmania infantum populations from southeastern Europe revealed by microsatellite analysis. Parasites & vectors. 2013;6:342. doi:10.1186/1756-3305-6-342.

[67] Kuhls K, Alam MZ, Cupolillo E, Ferreira GE, Mauricio IL, Oddone R et al. Comparative microsatellite typing of new world leishmania infantum reveals low heterogenei-

ty among populations and its recent old world origin. PLoS neglected tropical diseases. 2011;5(6):e1155. doi:10.1371/journal.pntd.0001155.

[68] Su C, Zhang X, Dubey JP. Genotyping of Toxoplasma gondii by multilocus PCR-RFLP markers: a high resolution and simple method for identification of parasites. Int J Parasitol. 2006;36(7):841-8. doi:10.1016/j.ijpara.2006.03.003.

[69] Herrmann DC, Maksimov P, Hotop A, Gross U, Daubener W, Liesenfeld O et al. Genotyping of samples from German patients with ocular, cerebral and systemic toxoplasmosis reveals a predominance of Toxoplasma gondii type II. International journal of medical microbiology : IJMM. 2014;304(7):911-6. doi:10.1016/j.ijmm.2014.06.008.

[70] Rajendran C, Su C, Dubey JP. Molecular genotyping of Toxoplasma gondii from Central and South America revealed high diversity within and between populations. Infection, genetics and evolution : journal of molecular epidemiology and evolutionary genetics in infectious diseases. 2012;12(2):359-68. doi:10.1016/j.meegid.2011.12.010.

[71] Cabral AD, Gama AR, Sodre MM, Savani ES, Galvao-Dias MA, Jordao LR et al. First isolation and genotyping of Toxoplasma gondii from bats (Mammalia: Chiroptera). Veterinary parasitology. 2013;193(1-3):100-4. doi:10.1016/j.vetpar.2012.11.015.

[72] Messenger LA, Yeo M, Lewis MD, Llewellyn MS, Miles MA. Molecular genotyping of Trypanosoma cruzi for lineage assignment and population genetics. Methods Mol Biol. 2015;1201:297-337. doi:10.1007/978-1-4939-1438-8_19.

[73] Marcili A, Lima L, Cavazzana M, Junqueira AC, Veludo HH, Maia Da Silva F et al. A new genotype of Trypanosoma cruzi associated with bats evidenced by phylogenetic analyses using SSU rDNA, cytochrome b and Histone H2B genes and genotyping based on ITS1 rDNA. Parasitology. 2009;136(6):641-55. doi:10.1017/S0031182009005861.

[74] Martins LP, Marcili A, Castanho RE, Therezo AL, de Oliveira JC, Suzuki RB et al. Rural Triatoma rubrovaria from southern Brazil harbors Trypanosoma cruzi of lineage IIc. The American journal of tropical medicine and hygiene. 2008;79(3):427-34.

[75] Du X, Dong L, Lan Y, Peng Y, Wu A, Zhang Y et al. Mapping of H3N2 influenza antigenic evolution in China reveals a strategy for vaccine strain recommendation. Nature communications. 2012;3:709. doi:10.1038/ncomms1710.

[76] Rico-Hesse R. Dengue virus markers of virulence and pathogenicity. Future virology. 2009;4(6):581. doi:10.2217/fvl.09.51.

Chapter 5

Nucleic Acids — The Use of Nucleic Acid Testing in Molecular Diagnostics

Gabrielle Heilek

Additional information is available at the end of the chapter

http://dx.doi.org/10.5772/61487

Abstract

In 1989 Roche entered into an agreement with Cetus to develop diagnostic applications for the novel technique polymerase chain reaction (PCR). A new area of molecular diagnostics began and genes and pathogen genomes have been used to diagnose disease since that point. Automated laboratory platforms were created to facilitate the workflow and allow for accurate and precise processing of patient blood samples in a highly streamlined manner. In this chapter the use of nucleic acids in molecular diagnostics will be described and their application to important human diseases. Examples are discussed with respect to which nucleic acid marker has provided strong clinical utility and impact to healthcare.

Keywords: Real Time-PCR, molecular diagnostic, HIV-1, human genetic testing

1. Introduction

In this chapter, the use of nucleic acids in molecular diagnostic testing will be described. Detailed disease area examples will be discussed to illustrate technical capabilities as well as the medical relevance of such testing.

Polymerase chain reaction (PCR) was invented by Kary B. Mullis in the 1980s. Fundamentally, PCR is a cyclic process designed to specifically replicate (amplify) nucleic acid sequences from as little as one to a few strands of DNA. The target DNA is heated to separate double-stranded DNA sequences; short oligonucleotide "primers" that define the portion of the genome to be replicated bind to the target DNA. The primers are extended by a DNA polymerase making a copy of the target DNA. After multiple cycles in which the concentration of the replicated target DNA increases exponentially, the amplified product (amplicon) can be visualized by gel electrophoresis or measured by detection of labeled PCR product (amplicon) by incubation with additional reagents to produce color or by fluorescent probe detection. This technique improved the ability to diagnose a number of diseases by enabling identification of many human pathogens that had previously been difficult to detect due to their low concentration in the sample. With the addition of a reverse transcription step to the original PCR process, RNA could be converted to cDNA and then replicated with the PCR process. Thus, the utility of the technique was broadened to detect RNA viruses and eukaryotic mRNA.

A key technical improvement was introduced by Higuchi et al.[4, 5] who developed real-time polymerase chain reaction (RT-PCR), which follows the kinetics of the PCR and detects PCR products during the process of amplification (Figure 1). With RT- PCR, accurate and reproducible quantitation of pathogen concentration could be incorporated into the amplification process. RT-PCR is used to monitor a pathogen's kinetic replication processes over time, and measurement of viral load is now widely employed to monitor the success of treatment of viral and other diseases.

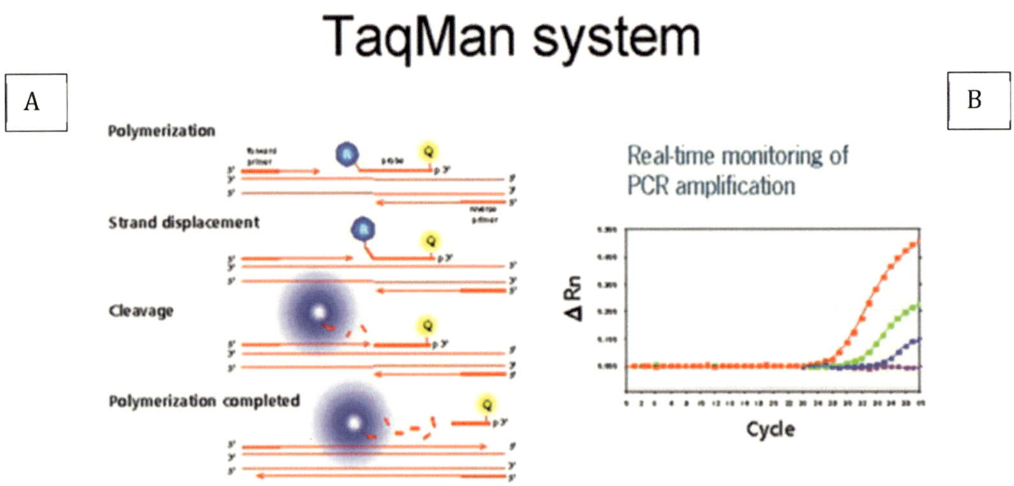

Figure 1. Principles of Real-Time PCR. (A) During the polymerization step, the template is amplified by primers supplied in the reaction mix. The amplicon allows for annealing of sequence-specific, labeled probes. As a new strand is synthesized, the probes will be displaced, the label cleaved off, and a fluorescent signal proportional to the amount of the cleaved probe is generated. (B) Fluorescence is measured and recorded at each cycle of PCR. Cycle threshold (Ct) is defined as the fractional PCR cycle number in which the sample fluorescence signal reaches a level above an assigned fluorescence threshold. The Ct value indicates the beginning of the exponential amplification of the template DNA or RNA and is proportional to the concentration of the sample.[6]

Clinical microbiology was one of the first fields to adopt PCR and, later, RT-PCR, due to the sensitivity and specificity of the technique for detecting nucleic acids of pathogenic microorganisms.

Perkin-Elmer developed the first thermal cycler instrument in December 1985. The first commercial *in vitro* PCR diagnostic products were created when the California company, Cetus, entered into a partnership with Kodak in February 1986.[7] The first reagent kit, the "Gene-Amp PCR reagent kit" and the thermal cycler were commercially available in November 1987.[8]

http://dx.doi.org/10.5772/61487

In January 1989, it was announced that Roche had entered into an agreement with Cetus to develop diagnostic applications for PCR. A new area of molecular diagnostics began using PCR to detect genes and pathogen genomes to diagnose diseases since that time.[7]

At Cetus, it was decided in the late 1980s that the forensic applications of PCR represented a stand-alone business that could be operated in-house, and, therefore, the applications were not sold to a partner.[7] In 1990, the first forensic PCR kit, developed by Cetus, was sold by Perkin-Elmer and became useful for identification of individual humans. The nucleic acid, DNA, became a mainstay of the justice system in 1997 when the FBI announced the selection of 13 short tandem repeat (STR) DNA loci to constitute the core of a national database—Combined DNA Index System (CODIS). By the time a review was published in 2006,[9] 5 million profiles of individuals existed in CODIS. By 2003, almost 1 million samples were being processed annually using core STR loci as part of parentage testing.[10]

Over the past decades, RT-PCR technology has continued to develop, optimize, and expand in the clinical laboratory for the identification, detection, and quantitation of a variety of pathogen uses. Automated instrument platforms were created to facilitate the workflow and allow for accurate and precise processing of patient samples in a highly automated manner.

2. Nucleic acid detection in molecular diagnostics

For molecular diagnostic purposes, since each microbe has a unique complement of DNA (or RNA, for many viral pathogens), nucleic acids are the ideal molecular fingerprint aiding identification. Particularly useful are RT-PCR and PCR, with their enzyme-driven processes for amplifying RNA/DNA *in vitro*, to analyze levels of microbial DNA in clinical samples for which other detection methods require higher concentrations, or are too time-consuming or cumbersome to detect.

2.1. The importance of nucleic acid measurement for HIV-1

With the emergence of the global HIV epidemic in the 1980s, it became evident that following the viral kinetics in infected subjects can aid significantly in understanding the progression of HIV-1 infection to stage 3, i.e., the disease of AIDS. Before viral load tests, many researchers believed that HIV-1 infection underwent dormant periods. Viral load tests showed that HIV-1 replication in the human body is a continual and gradually progressive process and that the viral replication is always active.[11] A typical pattern of HIV-1 infection is shown in Figure 2.

With the development of the first antiretroviral (ARV) agents (such as AZT, zidovudine) and later, the establishment of highly active antiretroviral therapy (HAART), understanding the kinetics of suppression of viral replication and detection of antiviral resistance became a major focal point in guiding and caring for patients. Viral load monitoring was the first direct approach to personalized healthcare, determining the activity of ARV medicines in an individual patient at specific time points.

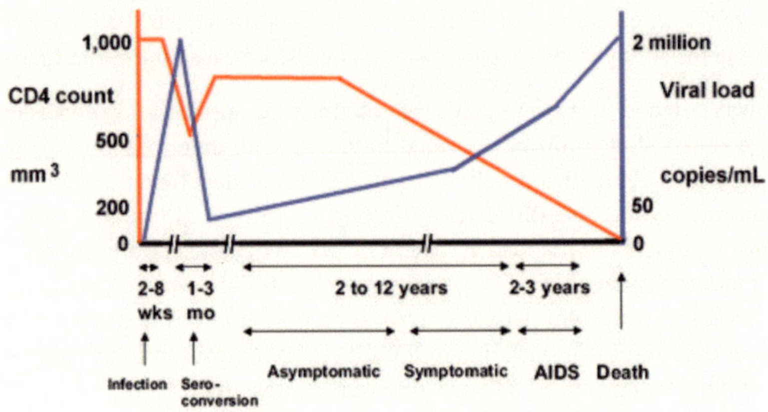

Figure 2. Viral Kinetics of HIV-1 Infection. After HIV-1 infection, during the acute disease phase, viral load is high, followed by a strong CD4 cell decline. After seroconversion and establishment of chronic infection, the viral load reaches a viral set point phase (at approximately 14 weeks following infection) from which it continues to rise as the CD4 cell count declines over several years. A CD4 cell count of fewer than 200 cells/mm^3 is one of the qualifications for a diagnosis of stage 3 infection (AIDS). Source: http://i-base.info/ttfa/section-2/214-how-cd4-and-viral-load-are-related/[12]

Suppressing the HIV-1 viral load to undetectable levels (<50 copies per mL) is the primary goal of HAART.[13] This level of suppression should be achieved by 24 weeks after starting combination therapy. HIV-1 viral load is the most important predictor of response to treatment with HAART.[15] Failure of HAART to adequately suppress viral load is termed virologic failure. Levels of HIV-1 RNA higher than 200 copies per mL are considered virologic failure, and should prompt further testing for potential viral resistance.

In 1992, to aid in this therapeutic approach the first commercially available PCR-based diagnostics were marketed. The AMPLICOR CT (*Chlamydia trachomatis*) Test and the AMPLICOR HIV-1 MONITOR Test were the first PCR-based molecular diagnostic tests. During the first half of the 1990s, the sensitivity of these early commercial products was moderate. By 1996 to 1997, through technical improvements, the next generation of tests could detect and measure viral loads as low as 400 or 500 copies/mL. Since 1998, most tests used routinely in clinical practice accurately detect and measure HIV-1 RNA as low as 40 or 50 copies/mL.[18] For academic research purposes, several groups have described an ultrasensitive or single copy assay that can detect 5 copies/mL or even 1 copy/mL.[19]

In addition to RT-PCR, a range of other nucleic acid-based techniques are also employed to measure HIV-1 RNA viral load, such as branched DNA (bDNA) assay[20] and nucleic acid sequence based amplification (NASBA).[21] NASBA was developed in 1991 by J. Compton, who defined it as "a primer-dependent technology that can be used for the continuous amplification of nucleic acids in a single mixture at one temperature."[22]

These techniques were also employed to study the relative effectiveness of ARV drugs in clinical trials during the very active HIV drug discovery decades from 1990 to 2010.[23]

http://dx.doi.org/10.5772/61487

(A) bDNA

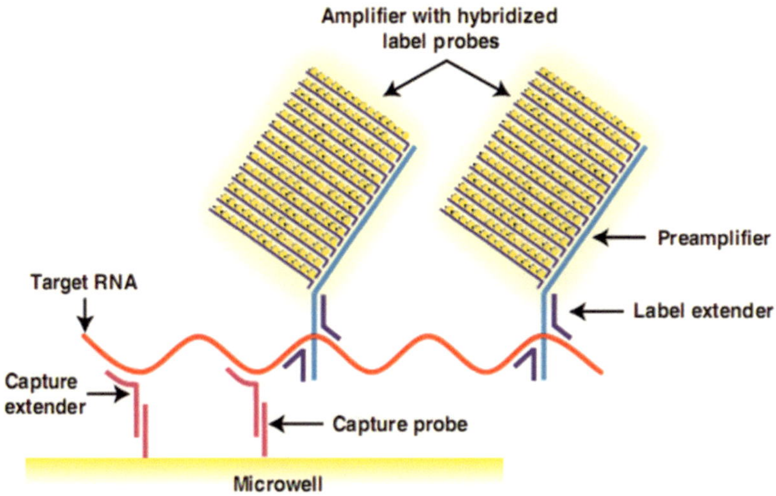

(B) NASBA

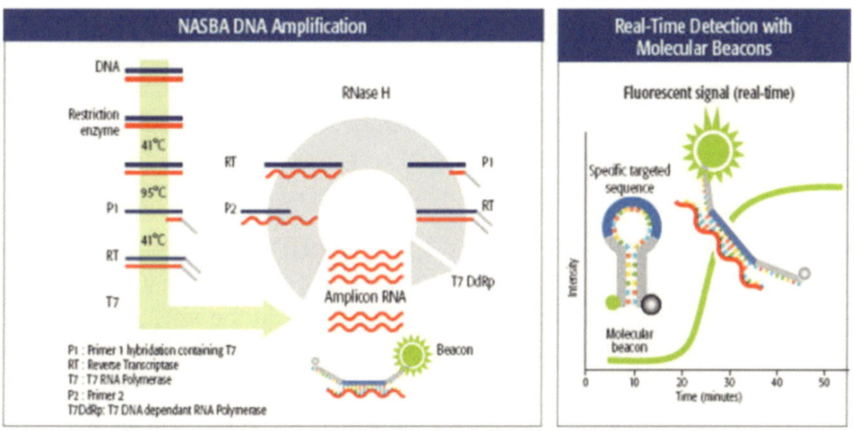

Figure 3. Principle of bDNA and NASBA. (A) Target RNA is captured with a bifunctional Capture Extender oligonucleotide probe that hybridizes to the target molecule and a Capture Probe that is covalently attached to a substrate (e.g., a microtitre plate well or a bead). A Signal Amplification complex (Preamplifier and Amplifier with labeled probes) containing a number of alkaline phosphatase enzymes is then hybridized to the target molecule via a Label Extender probe. Source: http://www.diacarta.com/article.php?id=38[24] (B) NASBA works as follows: An RNA template is added to the reaction mixture and reverse transcriptase synthesizes the opposite, complementary DNA strand. RNAse H destroys the RNA template from the DNA–RNA complex (RNAse H only destroys RNA in RNA–DNA hybrids, but not single-stranded RNA). A second primer attaches to the 5' end of the DNA strand. Reverse transcriptase again synthesizes another DNA strand from the attached primer, resulting in double-stranded DNA. T7 RNA polymerase continuously produces complementary RNA strands off this template which results in amplification. Finally, a molecular beacon is employed to detect the amplified product and allow for quantitation. Source: http://www.biomerieux.com.co/servlet/srt/bio/colombia/dynPage?open=CLM_CLN_PRD&doc=CLM_CLN_PRD_G_PRD_CLN_87&pubparams.sform=3&lang=es_co[25]

Various nucleic acid detection techniques, discussed above, are employed for the detection and quantitation of HIV-1 infection. Molecular diagnostics may carry, in general, a larger cost burden than other laboratory techniques that detect pathogens via shed surface proteins or antibodies in human serum. However, the speed, specificity, and sensitivity of molecular testing offers a number of advantages over the more "traditional" methods that it has replaced, such as culture, which is slow and labor-intensive, or hybridization or similar techniques that are often imprecise, insensitive, or for which the interpretation of results is often subjective. Molecular methods also offer advantages over measuring antigens, such as p24, or disease markers, such as CD4 cells, in people infected with HIV-1. Early monitoring of the status of patients with HIV-1 infection used the CD4 cell count to determine progression of disease, and many resource-limited settings still employ this technique today. CD4 cells are the white blood T-cells that are specifically targeted by HIV due to their surface receptor repertoire and depleted as infection progresses. The CD4 cell count provides a measure of the immune function of the human host and is a late marker of disease progression. The measurement is used in establishing thresholds for the initiation and discontinuation of opportunistic infection (OI) prophylaxis and in assessing the urgency to initiate HAART. It is recommended that ARV therapy be initiated when the CD4 cell count falls below 200–350 cells/mm^3, depending on the availability of ARV medicines in a given country.

Measurement of CD4 cells using the current technology is imprecise. Since certain standard-of-care recommendations, such as initiation of prophylaxis against *Pneumocystis carinii* pneumonia (an OI common in HIV-1 patients), have been made, treatment may be based on a single CD4 cell count, and CD4 measurement error may have important clinical consequences. Often the use of confirmatory tests is recommended and both tests need to be below a certain threshold limit.[27] Therefore, additional cost is incurred by the confirmatory testing, and the advantage of using the more inexpensive CD4 cell count test is lost.

After initiation of ARV therapy, due to suppression of the HIV-1 viral load, the immune system is allowed to recover and the CD4 cell count increases. For most patients on therapy, an adequate response is defined as an increase in CD4 count in the range of 50–150 cells/mm^3 during the first year of HAART, generally with an accelerated response in the first 3 months of treatment. The CD4 count response to HAART varies widely, but a poor CD4 response in a patient with viral suppression is rarely an indication for modifying an ARV regimen. In patients with consistently suppressed viral loads who have already experienced HAART-related immune reconstitution, the CD4 count provides only limited information.[28]

A second biomarker used in HIV-1 laboratory testing is the viral core protein p24. This biomarker can be measured in the patient's blood in early acute infection, often before antibodies to the viral onslaught are detectable. A negative result for the antigen does not rule out infection, because the test lacks exquisite sensitivity; i.e., the test should not be used to verify noninfection. Antigen detection signals infection, however, and positive results in seronegative individuals can be an effective, although not cost-effective, means to identify early infection. The p24 antigen test can be of value in blood screening, for identification of acute infection, for monitoring infection, and to assist in the diagnosis of infection in the newborn. It has been used for detecting early infection in rape cases, for identifying infection

http://dx.doi.org/10.5772/61487

after occupational exposure, and for assisting in the resolution of indeterminate Western blot results.[29]

As both CD4 cell count and p24 have caveats briefly discussed above, HIV-1 RNA viral load analysis by nucleic acid testing has, in many clinical situations, replaced less predictive methods of measurement of these biomarkers.

2.2. Other therapeutic areas

2.2.1. Microbiology and infectious diseases

Real-time PCR revolutionized the means by which clinical laboratories identify human pathogens. It is estimated that <1% of bacteria present on earth have been described using cultivation technology.[30] Additionally, various pathogens, particularly mycobacteria and fungi, require prolonged periods of cultivation, necessitating administration of empiric antimicrobial therapy while a laboratory result is awaited. Due to the limitations of cultivation technology, PCR amplification and sequencing-based methods are able to also reveal novel microbes associated with human diseases. Hence, cultivation-independent methods offer a potential for rapid diagnosis, thus preventing antibiotic selection pressure and emergence of resistant pathogen infections. Additionally, molecular testing is able to identify hazardous microbes without risk to laboratory staff as well as speed isolation of a given patient harboring highly infectious pathogens into a quarantine setting.

As opposed to monitoring during care of chronic viral infections, such as HIV-1, HBV, and HCV, most tests in the microbiology diagnostic assay repertoire are qualitative, aimed at the detection of the pathogens, which then warrant follow-up evaluation.

Several reviews of real-time PCR in clinical microbiology have been published; among others, the review by Espy et al. in 2006 is a comprehensive guide. Important pathogens diagnosed with molecular testing at the time of publication[32] of this chapter were agents for disease areas such as:

- Respiratory infections, such as adenovirus, Mycoplasma pneumoniae, Mycobacterium tuberculosis, Legionella spp, and Streptococcus pneumonia
- Genitourinary/sexually transmitted infections with PCR assays for C. trichomatis, Neisseria gonorrhoeae, Mycoplasma genitalium, and human papillomavirus
- Central nervous system infections dominated by herpes simplex, varicella zoster, and West Nile Virus (WNV)
- Gastrointestinal infections with, most notably, Clostridium difficile

In recent years, infectious disease surveillance and monitoring of antibiotic resistance has also been added to PCR-based molecular diagnostic tests, such as detection of gram-negative bacilli and vancomycin-resistant enterococcus species.

Finally, the host of viral pathogens causing human disease are generally identified, quantitated, and managed via PCR-based laboratory tests. Important examples are diagnosis and management of hepatitis B and C, herpes virus family infections, and influenza epidemic outbreaks.

2.2.2. *Blood screening*

Annually, millions of people worldwide receive blood transfusions or blood-derived products. Around the world, more than 92 million blood donations are collected every year.[33] From these, a single whole-blood donation can be transfused in up to three people, and blood-derived products from a single donation may be given to hundreds of patients.[34, 35] Although testing and policy decisions have combined to make blood supplies in many countries among the safest in the world, there still exists some risk of transfusion-transmitted infection (TTI) with blood-borne diseases (e.g., HIV, hepatitis, WNV). Laboratory screening of donated blood and blood products for infectious diseases is a key safety measure in protecting patients and preventing the spread of serious diseases.

Nucleic acid testing (NAT) by PCR- or transcription-mediated amplification (TMA) technology detects the presence of viral infection by directly testing for viral nucleic acids and can be used to screen whole blood and plasma samples. Commonly used NAT assays detect HIV-1 RNA, HCV RNA, HBV DNA, and WNV RNA.[38]

NAT technology has revolutionized the ability of blood centers to efficiently test for and reduce infusions of potentially infectious blood units while continuing to ensure on-time availability of blood and blood products for patients. The global trend toward adopting this technology clearly demonstrates its effectiveness for increasing the safety of blood supplies.

2.2.3. *Human genetics—Testing via nucleic acid markers*

Besides the exploration of human pathogen diagnostics, molecular testing has been employed to identify a myriad of human host markers predominately via DNA found in any human cell.

2.2.3.1. *Prenatal diagnosis*

Prenatal diagnosis employs a variety of techniques to determine the health and condition of an unborn fetus. There are three purposes of prenatal diagnosis: (1) to enable timely medical or surgical treatment of a condition before or after birth, (2) to give the parents the chance to abort a fetus with the diagnosed condition, and (3) to give parents the chance to prepare psychologically, socially, financially, and medically for a baby with a health problem or disability or for the likelihood of a stillbirth.

Congenital anomalies account for 276,000 perinatal deaths by pregnancy Week 4 annually on a global basis. The aim of prenatal screening is to detect birth defects, such as neural tube defects; chromosome abnormalities (e.g., Down Syndrome, fragile X syndrome); and genetic disorders and other conditions (e.g., spina bifida, cleft palate, Tay Sachs disease, sickle cell anemia, thalassemia, cystic fibrosis, and muscular dystrophy). Screening can also be used for prenatal sex discernment.

There is a variety of noninvasive and invasive techniques available for prenatal diagnosis. Each should be applied only during specific time periods of a pregnancy for greatest utility.

Traditionally, amniocentesis, performed at pregnancy Weeks 14–20, was employed to sample the amniotic fluid, which contains fetal cells, for analysis of chromosomal defects. Risks with amniocentesis are uncommon, but include fetal loss. The increased risk for fetal mortality following amniocentesis is about 0.5% above what would normally be expected. Collected embryonic cells from the amniotic sac need to be cultured for the chromosomal analysis. This process is cumbersome, carried out in specialized laboratories only, and requires a time period of 1–2 weeks, including transport of the sample.

Similarly, chorionic villi sampling can provide information about the fetus' health and development status as early as at 10 weeks of pregnancy. Miscarriage rates are higher in this procedure compared to amniocentesis, up to 1.9%. Test results are obtained within 2 weeks and require specialized laboratories and culturing techniques.

In recent years, analysis of cell-free DNA shed from fetal cells in the maternal blood has become a molecular technique to investigate congenital defects as early as pregnancy Weeks 9–10. High-throughput shotgun sequencing of the plasma of pregnant women results in obtaining about 5 million sequence tags per patient sample. Using this technology, in 2008, Fan et al. were able to identify aneuploid pregnancies, with trisomy detected at gestational ages as early as ~10 weeks. Shotgun sequencing is carried out on a next-generation sequencing platform such as Illumina. In 2010, Chiu et al. studied 753 pregnant females using a 2-plex massively parallel maternal plasma DNA sequencing, and trisomy was diagnosed with z-score greater than 3.[43] The test demonstrated 100% sensitivity, 97.9% specificity, positive predictive value of 96.6%, and negative predictive value of 100%.

The main advantages of these protocols are that they can be used earlier than the current prenatal testing protocols and, unlike current protocols, that there is no risk of spontaneous abortion. Noninvasive prenatal diagnosis (NIPD) has been implemented in the United Kingdom (UK) and parts of the United States (US).

2.2.3.2. Inherited diseases

Carrier screening, testing of parents in preparation for pregnancy, is used to identify genetic mutations that could cause serious inherited disorders. Some of the more common disorders for which screening is done are cystic fibrosis, sickle cell disease, thalassemia, and Tay-Sachs disease. These disorders are recessive, which means that a person must inherit a defective gene from each parent to have the disease. If both parents are carriers of a disorder, the child will have a one-in-four chance of inheriting one defective gene from each of the parents and having the disorder. This type of testing is offered to individuals who have a family history of a genetic disorder and to individuals in certain ethnic groups with an increased risk of specific genetic conditions. For the testing procedure, venous blood is collected and sent to specialized laboratories. There, the DNA contained in the human blood cells is amplified via PCR and, for example, a next-generation sequencing platform is utilized to investigate the genotype of a set of genes in a cost-efficient manner.

Newborn screening is used just after birth to identify genetic disorders that can be treated early in life. Early detection, diagnosis, and intervention can prevent death or disability and enable children to reach their full potential. Each year, millions of babies in the US are routinely screened, using molecular tests performed on a few drops of blood obtained from their heels, for certain genetic, endocrine, and metabolic disorders, and are also tested for hearing loss prior to discharge from a hospital or birthing center. All states currently test infants for phenylketonuria (a genetic disorder that, if left untreated, causes intellectual disability) and congenital hypothyroidism (a disorder of the thyroid gland).

The expansion of the screening panel to approximately 30 heritable metabolic conditions occurred from 1997 to 2007 with the introduction of tandem mass spectrometry (MS/MS), a technology that detects multiple disease biomarkers simultaneously in a single specimen. This technique employs the screening of blood spots for inborn errors of metabolism by electrospray MS/MS with a microplate batch process and a computer algorithm for automated flagging of abnormal profiles. More recently, other markers, based on nucleic acid analysis of the newborn genetic makeup,[48] such as sickle cell disease, alpha-1-antitrypsin deficiency, and Factor V Leiden, have been added.

2.2.3.3. Cancer markers

Cervical cancer is the 7th most common cause of cancer death in Europe for females, and the 15th most common cause of cancer death overall. According to currently available US Centers for Disease Control (CDC) Fast Stats [49] cervical cancer mortality in the US in 2010 was ~4,000 or ~2.5 deaths per 100,000 females.

The global statistics provided by Cancer Research UK are far more saddening. Worldwide, there were more than ~275,000 deaths from cervical cancer in 2010 that accounted for ~10% of female cancer deaths.

The Papanicolaou test—aka Pap test, Pap smear, cervical smear, or smear test—was historically the method of cervical screening used to detect potentially precancerous and cancerous cells in the endocervical canal of the female reproductive system. Atypical findings were followed with more sensitive diagnostic procedures, and, if warranted, interventions that aimed to prevent progression to cervical cancer.

In March 2014, the FDA's Medical Devices Advisory Committee Microbiology Panel voted unanimously to approve the cobas® 4800 HPV Test (Roche Molecular Systems) and recommended that this real-time PCR HPV test replace the Pap smear as the first-line standard of care for cancer screening, another use of nucleic acid testing in molecular diagnostics.

Another wide-ranging use of molecular tests using PCR can be found in the disease area of colorectal cancers where tumor nucleic acids are analyzed for the presence of mutations or other markers. Historically, all colorectal cancers (CRCs) have been considered a single disease entity sharing the same cause, clinical characteristics, and treatment outcomes. However, through analysis of precursor lesions and hereditary forms of the disease, it has now become clear that CRC is a complex and heterogeneous disorder. Although microsatellite instability (MSI) testing has been used for more than a decade for identifying patients with Lynch

http://dx.doi.org/10.5772/61487

syndrome, with the recent growth in personalized cancer care, other molecular tests to identify the genetic makeup of individual cancers have become increasingly more important in making therapeutic decisions. Novel medicines in oncology and relevant biomarker tests are now often developed side-by-side. Current indications for standard-of-care molecular testing in colorectal carcinomas include identifying hereditary cancer syndromes, such as Lynch syndrome (also known as hereditary nonpolyposis colorectal cancer [HNPCC]), and testing for *KRAS* mutational status as a predictor of response to antiepidermal growth factor receptor (EGFR) agents such as cetuximab. In the case of Lynch syndrome, multiple mononucleotide markers are detected via a fluorescent multiplex PCR-based method. A tumor tissue specimen (with tumor cellularity of >20%) and normal tissue specimen are amplified using PCR for 5–7 microsatellite markers. Patterns of normal and tumor genotypes are compared for each marker and scored as MSI-High, MSI-Low, or MS-Stable. Analysis for somatic mutations in the V600E hot spot in the BRAF gene may be indicated for tumors that are scored as MSI-High or show loss of MLH1 expression, because this mutation has been found in sporadic MSI-High tumors but not in HNPCC-associated cancer. *KRAS* mutations have been convincingly associated in randomized clinical trials with poor response to cetuximab and panitumumab. Activating mutations in *KRAS* serve to isolate this signaling pathway from the effects of EGFR and render EGFR inhibition ineffective. Recent advances have shown that only tumors with wild-type *KRAS* show significant response to these agents. Accumulated data from both randomized and nonrandomized studies, reviewed by Jimeno et al., suggest that patients with CRCs whose tumors show *KRAS* mutations should not receive EGFR-targeting monocloncal antibody therapy. This led to the so-called codiagnostic assays with guidance language in both the test intended use information and drug package insert detailing use of the molecular test results for physicians and patients. This approach of diagnostic testing prior to prescription of costly and not always easy-to-tolerate medicines will dominate personalized healthcare in the future.

Currently, most assays can be performed on small quantities of formalin-fixed paraffin-embedded–derived tumor DNA. The pathologist must carefully select the tumor block to minimize dilution of tumor DNA by contaminating normal cells, such as fibroblasts, endothelial cells, and inflammatory cells; a target of at least 10% tumor cells is recommended for most assays.

Cancer research continues to focus on new molecular markers.[55] The integration of molecular markers into existing histomorphologic classifications in surgical pathology has already provided additional stratification for a more accurate prognosis. Furthermore, a molecular definition of cancer may often guide therapy and allow the monitoring of residual disease.

3. Conclusion and a future outlook

The introduction of nucleic acid testing into clinical laboratories has vastly improved detection of infections. Chronic viral infection can be treated with tests at hand that are adequate to inform the physician if the patient is responding, developing resistance, or being cured. The safety of the blood supply was dramatically improved on a global basis with the introduction of nucleic acid testing for blood-borne pathogens. Expectant parents can be informed of the genetic risks of a pregnancy and the inherited diseases for which a developing fetus or a newborn may be treated. Finally, today, patients diagnosed with cancer can experience a much more tailored approach to therapy, maximizing success and efficiency and minimizing costs to both themselves and the healthcare system.

The past decade has seen the number of commercial molecular tests used in practice increase fivefold. In 2013, 60% of the molecular diagnostics tests were sold by five companies: Roche, Becton Dickenson, Abbott, Hologic, and Qiagen.[56] However, in recent years the number of companies developing molecular tests has grown remarkably. Roughly, 350 companies are now active in development of molecular diagnostics,[56] highlighting the utility and importance of nucleic acid testing in healthcare today.

As molecular testing becomes more widely available and applicable to healthcare globally, it is not surprising that the next-wave nucleic acid testing will penetrate the markets in emerging and developing countries. For example, ARV regimens are becoming more widely available, including in sub-Saharan Africa to manage the large numbers of HIV-infected individuals, and state-of-the-art viral load testing will need to accompany the expansion of these regimens. It is a challenge to the manufacturers of nucleic acid tests to adapt technologies and platforms to resource-limited settings. The future of molecular testing may involve reduction in time to test result as well as reduction in assay and instrument complexity and number and training expertise of staff required to perform such assays.

Acknowledgements

The author would like to thank Alison Murray, Ann Butcher, and Sandra Ruhl for critical review of the chapter and Sandra Ruhl for technical help in the preparation of the manuscript.

Author details

Gabrielle Heilek*

Address all correspondence to: gabrielle.heilek@roche.com

Roche Molecular Systems, Pleasanton, CA, USA

References

[1] Mullis KB. The unusual origin of the polymerase chain reaction. *Sci Am* 1990;262(4): 56-61, 64-5.

[2] Saiki RK, Scharf S, Faloona F, Mullis KB, Horn GT, Erlich HA, et al. Enzymatic amplification of beta-globin genomic sequences and restriction site analysis for diagnosis of sickle cell anemia. *Science* 1985;230(4732):1350-4.

[3] Saiki RK, Gelfand DH, Stoffel S, Scharf SJ, Higuchi R, Horn GT, et al. Primer-directed enzymatic amplification of DNA with a thermostable DNA polymerase. *Science* 1988;239(4839):487-91.

[4] Higuchi R, Dollinger G, Walsh PS, Griffith R. Simultaneous amplification and detection of specific DNA sequences. *Biotechnology* (NY) 1992;10(4):413-7.

[5] Higuchi R, Fockler C, Dollinger G, Watson R. Kinetic PCR analysis: real-time monitoring of DNA amplification reactions. *Biotechnology* (N Y). 1993 Sep;11(9):1026-30.

[6] Arya M, Shergill IS, Williamson M, Gommersall L, Arya N, Patel HR. Basic principles of real-time quantitative PCR. *Expert Rev Mol Diagn* 2005;5(2):209-19.

[7] Fore J, Wiechers IR, Cook-Deegan R. The effects of business practices, licensing, and intellectual property on development and dissemination of the polymerase chain reaction: case study. *J Biomedl Discov Collab* 2006;1:7.

[8] Smithsonian Videohistory Program: History of the PCR, session one transcripts. 1992.

[9] Butler JM. Genetics and genomics of core short tandem repeat loci used in human identity testing. *J Forensic Sci* 2006 Mar;51(2):253-65.

[10] American Association of Bloodbanks: Annual report summary for testing in 2003 prepared by the parentage standards program unit. October 2004. http://www.aabb.org/About_the_AABB/Stds_and_Accred/ptannrpt03.pdf.

[11] HIV i-Base. Section 2.9: History of viral load tests. In *Treatment Training Manual*. London, United Kingdom. 23 July 2011. Available at http://i-base.info/ttfa/section-2/29-history-of-viral-load-tests/ Accessed 13 July 2015.

[12] HIV i-Base. Section 2.14 How CD4 and viral load are related. In *Treatment Training Manual*. London, United Kingdom. 23 July 2011. Available at http://i-base.info/ttfa/section-2/214-how-cd4-and-viral-load-are-related/ Accessed 13 July 2015.

[13] Tsibris AM, Hirsch, MS. Chapter 128: Antiretroviral Therapy for Human Immunodeficiency Virus Infection. In *Mandell, Douglas, and Bennett's Principles and Practice of Infectious Diseases, 7th Edition.* London, United Kingdom. Churchill Livingstone; 2009: 1833-55.

[14] Thompson MA, Aberg JA, Hoy JF, Telenti A, Benson C, Cahn P, et al. Antiretroviral treatment of adult HIV infection: 2012 recommendations of the International Antiviral Society–USA panel. *JAMA* 2012;308(4):387-402.

[15] Murray JS, Elashoff MR, Iacono-Connors LC, Cvetkovich TA, Struble KA. The use of plasma HIV RNA as a study endpoint in efficacy trials of antiretroviral drugs. *AIDS*. 1999 May 7;13(7):797-804.

[16] Panel on Antiretroviral Guidelines for Adults and Adolescents. Guidelines for the use of antiretroviral agents in HIV-1-infected adults and adolescents. Department of Health and Human Services. Available at http://www.aidsinfo.nih.gov/ContentFiles/AdultandAdolescentGL.pdf. Accessed 13 July 2015.

[17] Murray JS, Elashoff MR, Iacono-Connors LC, Cvetkovich TA, Struble KA. The use of plasma HIV RNA as a study endpoint in efficacy trials of antiretroviral drugs. *AIDS* 1999;13(7):797-804.

[18] Medscape. HIV-1 Viral Load Testing and Roche Diagnostics Assay Kits. Available at http://www.medscape.com/viewarticle/410196. Accessed 13 July 2015.

[19] Hilldorfer BB, Cillo AR, Besson GJ, Bedison MA, Mellors JW. New tools for quantifying HIV-1 reservoirs: plasma RNA single copy assays and beyond. *Curr HIV/AIDS Rep* 2012 Mar;9(1):91-100.

[20] Pachl C, Todd JA, Kern DG, Sheridan PJ, Fong SJ, Stempien M, Hoo B, Besemer D, Yeghiazarian T, Irvine B, et al. Rapid and precise quantification of HIV-1 RNA in plasma using a branched DNA signal amplification assay. *J Acquir Immune Defic Syndr Hum Retrovirol* 1995 Apr 15;8(5):446-54.

[21] Kievits T, van Gemen B, van Strijp D, Schukkink R, Dircks M, Adriaanse H, Malek L, Sooknanan R, Lens P. NASBA isothermal enzymatic in vitro nucleic acid amplification optimized for the diagnosis of HIV-1 infection. *J Virol Methods* 1991 Dec;35(3):273-86.

[22] Compton J. Nucleic acid sequence-based amplification. *Nature*. 1991 Mar 7;350(6313):91-2.

[23] Saag MS. Use of HIV viral load in clinical practice: back to the future. *Ann Intern Med* 1997;126(12):983-5.

[24] DiaCarta Precision Molecular Diagnostics. SuperbDNA™ Technology. Hayward, CA. Available at http://www.diacarta.com/article.php?id=38 Accessed 14 July 2015..

[25] bioMérieux Colombia S.A.S. NucliSENS EasyQ® MRSA, Bogota, Columbia. Available at http://www.biomerieux.com.co/servlet/srt/bio/colombia/dynPage?open=CLM_CLN_PRD&doc=CLM_CLN_PRD_G_PRD_CLN_87&pubparams.sform=3&lang=es_co Accessed 14 July 2015.

[26] Giorgi JV, Cheng HL, Margolick JB, Bauer KD, Ferbas J, Waxdal M, et al. Quality control in the flow cytometric measurement of T-lymphocyte subsets: the multicenter AIDS cohort study experience. The Multicenter AIDS Cohort Study Group. *Clin Immunol Immunopathol* 1990;55(2):173-86.

[27] Hoover DR, Graham NM, Chen B, Taylor JM, Phair J, Zhou SY, et al. Effect of CD4+ cell count measurement variability on staging HIV-1 infection. *J Acquir Immune Defic Syndr* 1992;5(8):794-802.

[28] Department of Health and Human Services USA. Federally approved HIV/AIDS medical practice guidelines. Available at https://aidsinfo.nih.gov/guidelines Accessed 14 July 2015.

[29] Constantine N. HIV Viral Antigen Assays in HIV InSite Knowledge Base University of California San Francisco. September 2001. Available at http://hivinsite.ucsf.edu/InSite?page=kb-02-02-02-02#S3X accessed 14 July 2015.

[30] Fredricks DN, Relman DA. Application of polymerase chain reaction to the diagnosis of infectious diseases. *Clin Infect Dis* 1999;29(3):475-86; quiz 487-8.

[31] Espy MJ, Uhl JR, Sloan LM, Buckwalter SP, Jones MF, Vetter EA, et al. Real-time PCR in clinical microbiology: applications for routine laboratory testing. *Clin Microbiol Rev* 2006;19(1):165-256.

[32] Mayo Clinic. Microbiology Test Catalogue. Rochester, MN. 2015. Available at http://www.mayomedicallaboratories.com/test-info/microbiology/ Accessed 14 July 2015.

[33] Global Blood Safety and Availability: Facts and Figures from the 2007 Blood Safety Survey. WHO Fact Sheet #279, November 2009.

[34] Piotrowicz-Theizen D, Schoeffter C. An example of traceability up to the patient. *STP Pharma Pratiques* 2004;14(5):476-481..

[35] Yu MW, Mason BL, Guo ZP, et al. Hepatitis-C transmission associated with intravenous immunoglobulins. *Lancet* 1995;345:1173-1174.

[36] Koppelman MH, Assal A, Chudy M, Torres P, de Villaescusa RG, Reesink HW, et al. Multicenter performance evaluation of a transcription-mediated amplification assay for screening of human immunodeficiency virus-1 RNA, hepatitis C virus RNA, and hepatitis B virus DNA in blood donations. *Transfusion* 2005;45(8):1258-66.

[37] Parida M, Posadas G, Inoue S, Hasebe F, Morita K. Real-time reverse transcription loop-mediated isothermal amplification for rapid detection of West Nile virus. *J Clin Microbiol* 2004;42(1):257-63.

[38] Strobl F. NAT in blood screening around the world. Medical Laboratory Observer. 2011. Available at http://www.mlo-online.com/articles/201104/nat-in-blood-screening-around-the-world.php accessed 15 July 2015.

[39] Liu L, Oza S, Hogan D, Perin J, Rudan I, Lawn JE, et al. Global, regional, and national causes of child mortality in 2000-13, with projections to inform post-2015 priorities: an updated systematic analysis. *Lancet* 2015;385(9966):430-40.

[40] Klatt EC. Prenatal Diagnosis. The Internet pathology laboratory for medical education. Mercer University School of Medicine. Savannah, GA. Available at http://

library.med.utah.edu/WebPath/TUTORIAL/PRENATAL/PRENATAL.html Accessed 15 July 2015.

[41] Tabor A, Vestergaard CH, Lidegaard O. Fetal loss rate after chorionic villus sampling and amniocentesis: an 11-year national registry study. *Ultrasound Obstet Gynecol* 2009;34(1):19-24.

[42] Fan HC, Blumenfeld YJ, Chitkara U, Hudgins L, Quake SR. Noninvasive diagnosis of fetal aneuploidy by shotgun sequencing DNA from maternal blood. *Proc Natl Acad Sci U S A* 2008;105(42):16266-71.

[43] Chiu RW, Lo YM. Pregnancy-associated microRNAs in maternal plasma: a channel for fetal-maternal communication? *Clin Chem* 2010;56(11):1656-7.

[44] Umbarger MA, Kennedy CJ, Saunders P, Breton B, Chennagiri N, Emhoff J, et al. Next-generation carrier screening. *Genet Med* 2014;16(2):132-40.

[45] Millington DS, Kodo N, Norwood DL, Roe CR. Tandem mass spectrometry: a new method for acylcarnitine profiling with potential for neonatal screening for inborn errors of metabolism. *J Inherit Metab Dis* 1990;13(3):321-4.

[46] Rashed MS, Bucknall MP, Little D, Awad A, Jacob M, Alamoudi M, et al. Screening blood spots for inborn errors of metabolism by electrospray tandem mass spectrometry with a microplate batch process and a computer algorithm for automated flagging of abnormal profiles. *Clin Chem* 1997;43(7):1129-41.

[47] Chace DH, Kalas TA, Naylor EW. The application of tandem mass spectrometry to neonatal screening for inherited disorders of intermediary metabolism. *Annu Rev Genomics Hum Genet* 2002;3:17-45.

[48] Dobrowolski SF, Banas RA, Naylor EW, Powdrill T, Thakkar D. DNA microarray technology for neonatal screening. *Acta Paediatr Suppl* 1999;88(432):61-4.

[49] Centers for Disease Control and Prevention. Cervical cancer mortality. Deaths: Final Data for 2013. Atlanta, GA. Available at http://www.cdc.gov/nchs/fastats/pap-tests.htm Accessed 16Jul2015.

[50] Ferlay J, Soerjomataram I, Ervik M, Dikshit R, Eser S, Mathers C, Rebelo M, Parkin DM, Forman D, Bray, F. GLOBOCAN 2012 v1.0, Cancer Incidence and Mortality Worldwide: IARC CancerBase No. 11 [Internet]. Lyon, France: International Agency for Research on Cancer; 2013. Available at http://globocan.iarc.fr Accessed on 16 July 2015.

[51] Ferlay J, Steliarova-Foucher E, Lortet-Tieulent J, Rosso S, Coebergh JW, Comber H, et al. Cancer incidence and mortality patterns in Europe: estimates for 40 countries in 2012. *Eur J Cancer* 2013;49(6):1374-403.

[52] Shi C, Washington K. Molecular testing in colorectal cancer: diagnosis of Lynch syndrome and personalized cancer medicine. *Am J Clin Pathol* 2012;137(6):847-59.

[53] Domingo E, Niessen RC, Oliveira C, Alhopuro P, Moutinho C, Espin E, et al. BRAF-V600E is not involved in the colorectal tumorigenesis of HNPCC in patients with functional MLH1 and MSH2 genes. *Oncogene* 2005;24(24):3995-8.

[54] Jimeno A, Messersmith WA, Hirsch FR, Franklin WA, Eckhardt SG. KRAS mutations and sensitivity to epidermal growth factor receptor inhibitors in colorectal cancer: practical application of patient selection. *J Clin Oncol* 2009;27(7):1130-6.

[55] Yarbro JW, Page DL, Fielding LP, Partridge EE, Murphy GP. American Joint Committee on Cancer prognostic factors consensus conference. *Cancer* 1999;86(11): 2436-46.

[56] Kalorama Information. Five Companies Dominate Molecular Diagnostics. Available at http://www.kaloramainformation.com/article/2014-02/Five-Companies-Dominate-Molecular-Diagnostics Accessed 16 July 2015.

Chapter 6

Application of Nucleic Acid Aptamers to Viral Detection and Inhibition

Ana Gabriela Leija-Montoya, María Luisa Benítez-Hess and Luis Marat Alvarez-Salas

Additional information is available at the end of the chapter

http://dx.doi.org/10.5772/61773

Abstract

Nucleic acid aptamers are small oligonucleotides that specifically bind to other molecules through noncovalent interactions that rely on complex tridimensional structural arrangements. Aptamers are generated through the iterative *in vitro* selection method called SELEX, resulting in specific binding against a wide variety of molecular targets including viruses. Because aptamers are obtained *in vitro* and can be synthetically produced, they have been envisioned as future diagnostic and therapeutic tools for human diseases including virus-borne pathologies. Aptamers have been isolated against a number of viruses including pandemic influenza virus, human papillomavirus and hepatitis C virus. Although aptamers have proven themselves as extremely sensitive detection tools triggering the development of affordable and highly diagnostic methods, their use as therapeutic moieties has been hampered by biostability, delivery and pharmacodynamical issues. Nevertheless, a new generation of chemically modified aptamers shows promise for the coming of age of protein-targeted noncatalytic oligonucleotides for the therapy of viral disease. The present review focuses on the most successful antiviral aptamers reported and includes a description of some of the novel methods developed for their use as diagnostic and therapeutic tools

Keywords: Aptamer, Oligonucleotides, Nucleic acids, RNA, DNA

1. Introduction

Nucleic acid aptamers are small single-stranded oligonucleotides capable of adopting complex tertiary structures that allow noncovalent interactions with other molecules. Because aptamers closely interact with their targets, their structural features are essential for highly specific binding. The term *aptamer* was coined in 1990 from the Latin "aptus" meaning "fitting" and the Greek "meros" meaning "particle" [1]. Aptamers are generated through the iterative *in vitro* selection method called SELEX (systematic evolution of ligands by exponential enrichment) and are raised against a wide variety of molecular targets ranging from ions and macromolecules to whole organisms, including viruses, bacteria, yeast and mammalian cells [2]. Although many modifications to the SELEX method have been established to include new technologies and improve selection [3], the basic steps of SELEX remain immutable.

The SELEX method involves three well-defined steps [4]: the start point is the production of a synthetic oligonucleotide combinatorial library or oligonucleotide pool containing a central randomized region (15–70 nt) flanked by anchor sequences to allow polymerase chain reaction (PCR) amplification. The aleatory nature of the central region results in the production of an enormous pool of diverse oligonucleotides with diverse structures, thus providing the conformational variability necessary to produce moieties with binding capabilities for a desired target. The oligonucleotide pool can be directly used for SELEX to generate single-stranded DNA (ssDNA) aptamers, or as *in vitro* transcription template to produce an RNA pool to isolate RNA aptamers. Next, a selection procedure is performed based on the interaction properties of the library with the intended target. Only a very small fraction of the oligonucleotide pool tends to interact with the target, satisfying the selection criteria. Oligonucleotides that bind the target (aptamers) are recovered while the nonbound are removed through different strategies according to the nature of the aptamer–ligand complex (size, affinity, electric charge, hydrophobicity, etc.). In the final step, the recovered aptamers are amplified by PCR in order to regenerate a library with less variability but more affinity to the target that will be used in the next selection cycle. RNA pools are amplified by reverse transcription-coupled PCR (RT-PCR) and subsequent *in vitro* transcription before starting the next cycle.

The iterative selection cycles produce aptamers with high binding affinity to the target. Usually, a few cycles are required to isolate aptamers (4–20 cycles), but the precise number of cycles necessary for the isolation of highly specific aptamers depends on the selection criteria, the nature of the target and the type of library used. After the last selection cycle, aptamers are cloned and sequenced to obtain information on the individual oligonucleotides, which can be further characterized based on its ability to bind the target. It is common to observe conserved sequences or structures among the selected aptamers; these are indicative of efficient selection and may represent domains required for interaction.

Aptamer specificity is based on three-dimensional arrangements of a small number of contact points between the aptamer and its target, so the aptamer can achieve high selectivity to discriminate between two highly related molecules (i.e. enantiomers), or minimal structural differences such as the presence or absence of methyl or hydroxyl groups. The molecular recognition specificity and affinity level achieved by aptamers is comparable or even better than those of antibodies. These features place aptamers as an emerging class of molecules on their own with a huge range of diagnostic and therapeutic applications plus several advantages over antibodies including:

- Isolation by an *in vitro* process not dependent on animal cells or *in vivo* conditions. Therefore, the properties of aptamers can change on demand, and isolation can be manipulated to obtain aptamers with desirable properties for diagnosis. In addition, it allows aptamer isolation against toxins or poorly immunogenic molecules.

- Production by chemical synthesis with accuracy and reproducibility, thus insuring mass production with high quality control standards.
- Aptamers can be reversibly denatured allowing conditional binding through simple temperature control.

Since the development of SELEX, aptamers have been isolated against a wide diversity of targets such as amino acids [5, 6], antibiotics [7], nucleotides [8], enzymes [9], growth factors [10], mammalian cells [11], bacteria [12] and parasites [13]. Nowadays, some aptamers have even reached therapeutic applications in the clinic [14]. Furthermore, the first RNA aptamer for therapeutic purposes in humans (pegaptanib sodium or Macugen®) was approved by United States Food and Drug Administration (FDA) in 2004, as treatment for age-related macular degeneration (AMD) [15].

2. Aptamers against viruses

Many aptamers have been isolated against whole viruses or viral proteins to detect or inhibit infection. Viruses such as human papillomavirus (HPV), human immunodeficiency virus-1 (HIV-1), hepatitis C virus (HCV), hepatitis B virus (HBV), severe acute respiratory syndrome coronavirus (SCoV), influenza virus, herpes simplex virus (HSV), Ebola virus, Rift Valley fever virus, dengue virus, human T cell leukemia virus type-1 (HTLV-1), Epstein–Barr virus and human cytomegalovirus (HCMV) have all been targeted with aptamers [16].

Aptamer isolation to inhibit viral infection can be performed by using purified molecules from the viral surface through canonical SELEX approaches, or by modified SELEX methods with the use of attenuated whole viral particles. The advantage of this last variant is the isolation of aptamers through binding to the native viral conformation. Moreover, a deep knowledge of the viral infection mechanisms or potential surface target molecules is not required to obtain antiviral or neutralizing aptamers that tightly bind infectious particles. On the other hand, this method does not disclose the sites that directly interact with the aptamers, so further studies are required to determine specific interactions useful for potential aptamer improvement.

Many efforts have been focused on the isolation of aptamers to detect and treat viral diseases relevant to public health such as AIDS, hepatitis, influenza and some cancers. Here, we summarize the successful application of aptamers selected against HIV-1, HPV, HCV, and influenza.

2.1. Aptamers against Rous Sarcoma Virus (RSV)

The first approach using whole viruses to isolate RNA aptamers without previous knowledge of the virion structural features was performed against Rous sarcoma virus (RSV), an avian retrovirus. Nineteen RNA aptamers were isolated from a canonical SELEX procedure and five of them were able to neutralize the virus infection [17]. These results immediately revealed the potential of aptamers isolated against viral surface epitopes leading to the development of nucleic acid aptamers as novel diagnostic or therapeutic tools, especially on human viral diseases requiring fast diagnostics (i.e. pandemic influenza or Ebola) or asymptomatic chronic viral-induced conditions such as acquired immunodeficiency syndrome (AIDS), hepatitis C or cervical cancer [18].

2.2. Aptamers against Human Immunodeficiency Virus type I (HIV-1)

HIV-1 is the etiologic agent of AIDS [19, 20]. Most anti-HIV-1 aptamers are directed to HIV-1 reverse transcriptase (RT), RNaseH, integrase, Tat, Gag, nucleocapsid, gp120 and the TAR-element RNA (Table 1). The HIV-1 RT is the enzyme responsible for transforming the viral genomic RNA into dsDNA and contains a domain with RNaseH activity. HIV-1 RT is also the main target of several therapies against AIDS. So far, about a dozen of ssDNA and RNA aptamers have been reported to inhibit the RT activity in cell cultures showing K_D in the range of 25 pM to 30 nM [18, 21].

Aptamer	Nature	Sequence	Randomized region	Target	Action	Structure	Ref
P5	RNA	GGGAGCUCAGAAUAAACG CUCAACGGCACAGGGGUU GUAUCCUCCGGGACGAAU UCGACAUGAGGCCCGGAU CCGGC	30 nt	Integrase	Inhibit interaction between integrase and viral DNA		Allen P, *et al.* 1995
A54	RNA	GGGAGCUCAGAAUAAACG CUCAAGUCAAUCAUCGAU GUCCUGUGCCCUAGGGCU UCGACAUGAGGCCCGGAU CCGGC	30 nt	Integrase	Inhibit interaction between integrase and viral DNA		Allen P, *et al.* 1995
93del	DNA	GGGGTGGGAGGAGGGT	80 nt	Integrase	Block integrase actvity in vitro	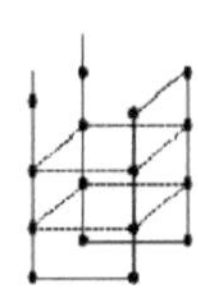	Phan AT, *et al.* 2004, De Soultrait VR. Et. al. 2002
112del	DNA	CGGGTGGGTGGGTGGT	80 nt	Integrase	Inhibton of HV-1 integrase	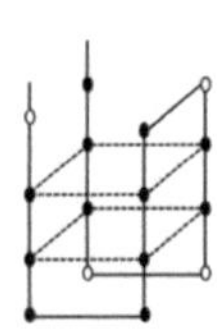	, De Soultrait VR. Et. al. 2002

Aptamer	Nature	Sequence	Randomized region	Target	Action	Structure	Ref
RNA tat	RNA	ACGAAGCUUGAUCCCGUU UGCCGGUCGAUCGCUUCG A	120 nt	TAR	Inhibition of Tat dependent trasns-activation transcription. Biosensor		Yamamoto R, *et al.* 2000
B40	RNA	TAATACGACTCACTATAGG GAGACAAGACTAGACGCT CAaTGTGGGCCACGCCCGA TTTTACGCTTTTACCCGCAC GCGATTGGTTTGTTTTCGA CATGGACTCACAACAGTTC CCTTTAGTGAGGGTTAATT	40 nt	Gp120	Neutralizaton of HIV-1 infectivity	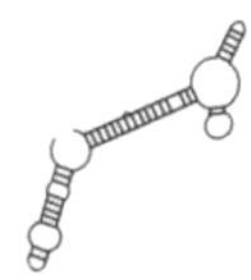	Khati M, *et al.* 2003. Dey AK, *et al.* 2005.
B40t77	RNA	TAATACGACTCACTATAGG GAGACAAGACTAGACGCT CAATGTGGCCACGCCCGAT TTTACGCTTTTACCGCACG CGATTGGTTTGTTTCCC	40 nt	Gp120	Neutralizaton of HIV-1 infectivity		Dey AK, *et al.* 2005. Cohen C, *et al*, 2008.

Table 1. Aptamers isolated against HIV proteins.

The HIV-1 integrase incorporates the viral DNA in the host genome [22]. RNA aptamers were isolated targeting HIV-1 integrase and classified into three groups according to their K_D (10 nM, 80 nM and 800 nM) [23]. Furthermore, DNA aptamers were also isolated targeting HIV-1 integrase by two different research groups, all with G-quadruplex structures and showing *in vitro* inhibitory activity [24, 25]. The characteristic structure of aptamers binding the HIV-1 integrase interacts within a channel of the tetrameric protein blocking catalytic amino acid residues essential for integrase function *in vitro* [26].

HIV-1 Tat protein regulates viral gene expression by interaction with the trans-activation responsive (TAR) elements within the long-terminal repeats (LTRs) [27]. Unlike the natural target of Tat (TAR-1 RNA), the isolated RNA aptamer (RNA^{Tat}) was highly specific to Tat and did not interact with other cellular factors. Moreover, RNA^{Tat} binds Tat protein over 100-fold higher than TAR-1 RNA and inhibited Tat function *in vitro* and *in vivo* [28, 29]. Based on these results, RNA^{Tat} was used in a preliminary study to develop a molecular beacon by flanking the 5′ and 3′ ends of the native aptamer stem-loop structure with a fluorophore and a quencher. In the absence of Tat, the quencher and fluorophore remain close to each other by the formation of the stem producing no signal. When the loop interacts with Tat, the complexed structure becomes more stable resulting in strand separation, thus holding apart the fluorophore and quencher allowing fluorescence [30].

Two other biosensors have been developed using RNA aptamers specific to HIV-1 Tat. These biosensors were created by immobilizing a biotinylated aptamer on a streptavidin layer over quartz crystals included in surface plasmon resonance (SPR) chips [31]. Another approach used a diamond field-effect transistor (FET) technique to detect Tat protein by RNA aptamers. Aptamer-FET is based on a gate potential shift generated by the presence of HIV-1 Tat bound to the RNA aptamer on a solid diamond surface. Efficient detection showed a potential use for aptamer-FET in clinical applications [32].

HIV-1 Rev is essential to regulate the splicing and shuttle the viral mRNA through their nuclear and export localization signals. Also, Rev interacts with viral mRNA through a cis-acting Rev-binding element (RBE) within a Rev-responsive element (RRE). RNA aptamers against Rev have been isolated using random libraries or by randomizing the RRE minimal binding sequence [33–35]. Randomized RRE produced aptamers with up to 16-fold tighter binding than the minimal wild-type RBE (wtRBE). The RBE was then substituted by these RNA aptamers and tested *in vivo* using a reporter system [35]. The aptamer substitutions showed a better response than wtRBE [36]. Another technique used to isolate RNA aptamers against Rev protein was the cross-linking SELEX consisting of a 5′-iodo uracil (5-IU)–modified RNA library, which is reactive under long-wavelength UV irradiation producing cross-links between 5-IU oligonucleotides and the protein target. This method resulted in the selection of highly specific aptamers capable of forming covalent bonds with HIV-1 Rev [37]. Some other efforts have focused on the inhibition of HIV-1 replication in human T cells using RNA aptamers as decoys to sequester Rev [38]. Rev decoys and ribozymes have been combined to increase the anti-HIV effect relative to independent ribozyme or decoy effects [39–41].

HIV-1 gp120 is a surface glycoprotein involved in the early stages of HIV-1 infection. The gp120 protein interacts with the human surface receptor CD4 producing conformational changes and further receptor interactions to allow HIV-1 entry into the host cell. Due to its importance on the onset of the viral infection, gp120 represents a potential target for the isolation of aptamers to block HIV-1 entry. Several RNA aptamers have been isolated to block gp120 and CD4 interaction neutralizing diverse subtypes of the virus [42]. Characterization of aptamer B40 showed high specificity to HIV-1 R5 strain and neutralization in human peripheral blood mononuclear cells [43, 44]. Additional analyses produced a shorter synthetic B40 derivative (UCLA1) able to inhibit entry of HIV-1 at the nanomolar range. Moreover, the aptamer showed synergistic effects with a gp41 fusion inhibitor (T20) and anti-CD4 binding site monoclonal antibody (IgG1b12), suggesting a potential use as adjuvant [45].

2′-Fluoride (2′-F) modified RNA aptamers selected to bind HIV-1_{Bal} gp120 and specifically internalized by cells expressing HIV-1_{Bal} gp120 were used to deliver anti-HIV siRNA into HIV-1–infected cells [46]. Two aptamer-siRNA chimeras were used: one covalent chimera presented a 2′-F-modified gp120 aptamer covalently attached to the sense strand of *tat/rev* siRNA and reduced the plasma viral load in a RAG-hu mouse model by suppressing HIV-1 replication and preventing CD4+ T cell decline. This effect was extended by several weeks beyond the last dose [47]. The second chimera consisted of a single aptamer with three different siRNAs targeting viral and cellular transcripts. The siRNA was linked to the aptamer by a bridge sequence of 16 nt that allowed complementary base pairing of one of the two siRNA strands to the aptamer. The aptamer–siRNA chimera showed a potent suppression of HIV-1 and protection from viral CD4+ T-cell depletion *in vivo*. In addition, the inhibitory effects were

http://dx.doi.org/10.5772/61773

also extended several weeks after the last injection, providing an attractive therapeutic approach to HIV-1 therapy [48].

2.2.1. *Aptamers against HPV*

HPVs are small DNA viruses that infect squamous epithelia inducing proliferative lesions ranging from benign warts to cancer. High-grade papillomavirus, especially types 16 and 18 (HPV-16 and HPV-18) are associated with cervical carcinoma, the second most common cancer affecting women worldwide. HPVs have a circular double-stranded DNA genome of approximately 8 kb that is organized into three regions: the upstream regulatory region (URR), the early region (E) and the late region (L). The URR contains several transcription factor binding sites to control gene expression, the early region encodes six genes (E1, E2, E4, E5, E6 and E7) involved in viral replication, transcription and cell transformation and the late region encodes the L1 and L2 capsid proteins which self-assemble to produce the virion [49].

Because preventive vaccines for HPV infection are only protective for *naive* individuals [50], several research groups have been developing nucleic acid–based aptamers targeting HPV proteins in order to inhibit the oncoproteins activity, block viral infection or identify the absence/presence of viral proteins as biomarkers to determine cell transformation or cancer progression (Table 2).

Aptamer	Nature	Sequence	Randomized region	Target	Function	Structure	Ref
F2	RNA	GGGAAUGGAUCCA CAUACUACGAAUA UUCAACAUUCGAG GUGGAUGCUACGA AUCAACUUCACUG CAGACUUGACGAA GCUU	30 nt	E6	Inhbition of E6-PDZ (Magil1) interaction and Induction of apoptosis in SiHa cells	ND	Belyaeva TA, *et al.* 2014
F4	RNA	GGGAAUGGAUCCA CAUACUACGAAAA CUCGUUUCGAGGU UCGAAACGUUGUA AAGCCGUUUCACU GCAGACUUGACGA AGCUU	30 nt	E6	Inhbition of E6-PDZ (Magil1) interaction and Induction of apoptosis in SiHa cells	ND	Belyaeva TA, *et al.* 2014
A2	RNA	GGGAAUGGAUCCA CAUCUACGAAUCC CUUCAUCAUUAAC CCGUCCACGCGCU UCACUGCAGACUU GACGAAGCUU	30 nt	E7	Inhibition of E7-pRb interaction and Induction of apoptosis in SiHA cells		Nicol C. *et al*, 2013

Aptamer	Nature	Sequence	Randomized region	Target	Function	Structure	Ref
G5a3N.4	RNA	GGGAGACCCAAGCCGAUUUAUUUUGUGCAGCUUUUGUUCCCUUUAGUGAGGGUUAAUU	15 nt	E7	E7 high affinity binding on HPV-positive cervical carcinoma cells	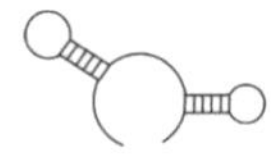	Toscano-Garibay JD, *et al.* 2011
Sc5-c3	RNA	GGGAACAAAAGCUGCACAGGUUACCCCCGCUUGGGUCUCCCUAUAGUGAGUCGUAUUA	15 nt	L1	High affinity binding of HPV VLPs in murire biofluids		Leija-Montoya AG, *et al.* 2014
C5	RNA	GGGAGGACGAUGCGGAAGCAUCAAGGGTGATCGTTTGACCCTCCCCAGACGACUCGCCCGA	30 nt	HPV-16 E6/E7-HTECs	Internalization in HPV-16 E6/E7 HTEC as mechanism to deliver therapeutc agents		Gourronc FA, *et al.* 2013
13	DNA	ATACCAGCTTATTCAATTGGGCACAGACGGAAGATGAGAATTGTGGGGCTTAGTATAGTGAGGTGCGTGTAGATAGTAAGTGCAATCT	52 nt	HF cell line	Detection of biomarkers lost in HPV-mediated cell transformation		Graham JC, *et al.* 2012
14	DNA	ATACCAGCTTATTCAATTGGGCGGGGAGTAGGGAGAGGGGTTTCCATCGGCGACAGAGGAGTTATGTGTGTAGATAGTAAGTGCAATCT	52 nt	HF cell line	Detection of biomarkers lost in HPV-mediated cell transformation	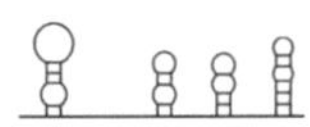	Graham JC, *et al.* 2012
20	DNA	ATACCAGCTTATTCAATTGGGGAGGGAGACACAGTCATGGAGCAGTTATTAGGGTGTACCGGGTGTAGTAGATAGTAAGTGCAATCT	52 nt	HF cell line	Detection of biomarkers lost in HPV-mediated cell transformation	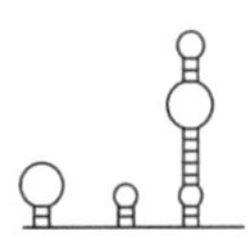	Graham JC, *et al.* 2012
28	DNA	ATACCAGCTTATTCAATTGGGGGACACGGAGGTGGTGGAA	52 nt	HF cell line	Detection of biomarkers lost in HPV-mediated cell transformation	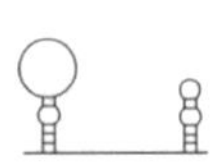	Graham JC, *et al.* 2012

Aptamer	Nature	Sequence	Randomized region	Target	Function	Structure	Ref
		AGGCTAAGATTTGA TGATGAGTAGTGTG GTAGATAGTAAGTG CAATCT					

Table 2. Aptamers isolated against HPV proteins.

The oncoproteins E6 and E7 are involved in cell immortalization and malignant transformation. E6 promotes the degradation of the tumor suppressor p53 [51], and E7 binds and destabilizes the cell cycle control protein pRb [52]. E6 and E7 have an important role in cancer progression, situating these oncoproteins as the principal potential targets to bind aptamers to block their oncogenic activity and cancer progression.

Several RNA aptamers were isolated against the PDZ-binding motif of the HPV-16 E6 oncoprotein, two of them were able to inhibit the interaction between E6 and proteins with PDZ domain (Magi 1) resulting in apoptosis. The aptamer interaction with PDZ domain was very specific and the interaction between E6 and p53 was not affected [53]. The same research group also isolated RNA aptamers against E7 oncoprotein that were able to disturb the E7–pRb interaction by targeting E7 for degradation and showed that one of them (A2) was able to inhibit cellular proliferation by inducing apoptosis in SiHa cervical carcinoma cells [54]. This effect was specific to HPV-16 transformed cells because it was not observed in HPV-free or HPV-18 cell lines [55]. Specific apoptosis induction of RNA aptamers targeting E6 and E7 oncoproteins suggests that these aptamers could have further applications in the future as therapeutic moieties.

A deeply characterized RNA aptamer targeting HPV-16 E7 oncoprotein named G5α3N.4 interacts with E7 through two stem-loop motifs in a clamp-like manner, suggesting a change in aptamer structure due to protein contact. The complex formation was observed exclusively in HPV-positive cervical carcinoma cells, suggesting that G5α3N.4 could be used to detect HPV infection and cervical cancer [56, 57].

The L1 protein is the main component of the HPV capsid. It is arranged in 72 capsomers, each consisting of five 55-kDa L1 monomers and a single 74-kDa L2 unit (theoretical 5:1 ratio). The L1 protein can self-assemble, forming virus-like particles (VLPs) that are structurally and immunologically similar to the infectious virions. HPV-16 L1 VLPs have been broadly used in HPV virology research, as delivery agents for epitopes or genes and to successfully produce prophylactic vaccines against HPV infection. The first RNA aptamer, targeting the L1 protein (Sc5-c3), was obtained using HPV-16 VLPs as targets [58]. Sc5-c3 structure consists of a hairpin structure with a 16-nt loop that directly binds VLPs with very low K_D (0.05 pM). This aptamer was able to specifically bind VLPs in complex protein mixtures (murine cervical washes), suggesting that Sc5-c3 may provide a potential diagnostic tool for active HPV infections and, with further refinement, could be used as a potential tool to inhibit viral infection [58].

Nucleic acid–based aptamers have been also isolated against whole HPV-infected cells. A cell-based SELEX protocol (cell-SELEX), was used to isolate RNA aptamers able to internalize into HPV-16 E6/E7 transformed human tonsillar epithelial cells (HTEC). This was the first report of aptamers that specifically internalize into HPV-16–transformed cells, providing a plausible mechanism to specifically deliver therapeutic agents into HPV-16–associated tumors [59]. Moreover, DNA aptamers have been isolated by a cell-SELEX modification for use with adherent cells (AC-SELEX). These aptamers recognize cell surface differences between HPV-transformed and nontumorigenic cell lines and one of them (Aptamer 14) was able to enter the cells independent of cell surface protein binding. These selected aptamers have potential to elucidate biomarkers for cellular changes associated to nontumorigenic phenotype in HPV-infected cells [60].

2.3. Aptamers against influenza virus

Influenza viruses are associated with most flu pandemics. They are enveloped RNA viruses of 80 to 120 nm diameter that infect the upper respiratory tract. The disease severity depends on the virus type: A, B or C. Influenza A virus infects birds and mammals, influenza B targets mainly humans and influenza C is less common than A or B but it also causes disease. Although the three virus types infect different hosts, it has been reported that all of them can infect humans and thus they have been the subject of several SELEX protocols.

2.3.1. Influenza A Virus (IAV)

The IAV genome comprises eight segments of linear RNA and two surface glycoproteins: hemagglutinin (HA) and neuraminidase (NA). These proteins are used to classify the IAV subtypes. Seventeen HA (H1–H17) and nine NA (N1–N9) variants have been identified and implicated on viral attachment, membrane fusion and viral entry to the host cell. Many aptamers have been isolated to bind HA and NA in order to inhibit and detect the viral infection, mainly H5N1, H9N2, H1N1 and H3N2 subtypes (Table 3).

Aptamer	Nature	Sequence	Randomized region	Target	Function	Structure	Ref
H3N2							
A22	DNA	AATTAACCCTCACTAA AGGGCTGAGTCTCAAA ACCGCAATACACTGGT TGTATGGTCGAATAAG TTAA	30 nt	HA (91-161)	Inhibition of viral infection		Jeon SH, *et al.* 2004
A21	DNA	AATTAACCCTCACTAA AGGGCGCTTATTTGTTC AGGTTGGGTCTTCCTAT TATGGTCGAATAAGTT AA	30 nt	HA (91-161)	Inhibition of viral infection		Jeon SH, *et al.* 2000

http://dx.doi.org/10.5772/61773

Aptamer	Nature	Sequence	Randomized region	Target	Function	Structure	Ref
P-30-10-16	RNA	GGGAGAAUUCCGACCAGAAGGGUUAGCAGUCGGCAUGCGGUACAGACAGACCUUUCCUCUCUCCUUCCUCUUCU	30 nt	A/Panama/2007/1999	Inhibition of viral infection. Discriminate between related H3N2	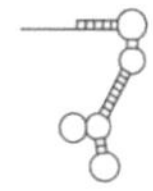	Gopinath SC, *et al.* 2006
H5N1							
10	DNA	GATTCAGTCGGACAGCGGGGTTCCCATGCGGATGTTATAAAGCAGTCGCTTATAAGGGATGGACGAATATCGTCTCCC	40 nt	HA	In vitro Inhibition of viral infection		Cheng C, et al.2008
2	DNA	GTGTGCATGGATAGCACGTAACGGTGTAGTAGATACGTGCGGGTAGGAAGAAAGGGAAATAGTTGTCCTGTTG	74 nt	HA and whole H5N1	H5N1 detection (QCM aptasensor)	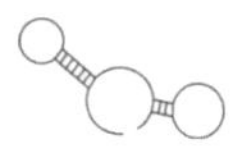	Wang R,and Li Y. 2013
H9N2							
A9	DNA	GCTGCAATACTCATGGACAGCCTCCTGGGGTCAGGCTCAGACATTGATAAAGCGACATCGGTCTGGAGTACGACCCTGAA	40 nt	HA	Inhibition of viral infection	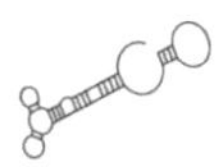	YueweiZhang, *et al.* 2015
B4	DNA	GCTGCAATACTCATGGACAGGGGCCGCGCCTGGTCGGTTGGGTGGGTGGCGCCCGGGACGGTCTGGAGTACGACCCTGAA	40 nt	HA	Inhibition of viral infection	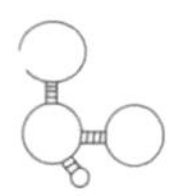	YueweiZhang, *et al* 2015
H5N1 AND H7N7							
8-3S	RNA	GGGCAACCGCUGGAACUUGAAGUCGGUAAUGCGAGCGGAAAGCCC	70 nt	HA	Discriminate IVA subtypes, inhibition of receptor binding	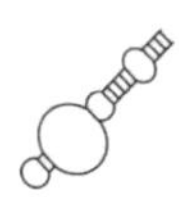	Suenaga E and Kumar PK, 2014.
H5N1, HIN1 AND H3N2							
RHA0006	DNA	GGGTTTGGGTTGGGTTGGGTTTTTGGGTTTGGGTTGGGTTGGGAAAAA	30 nt	rHA	IVA detection (ELAA)	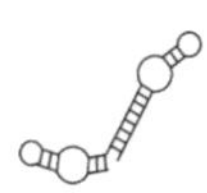	Shiratori I, *et al.* 2014
RHA0385	DNA	TTGGGGTTATTTTGGGAGGGCGGGGGTT	30 nt	rHA	IVA detection (ELAA)	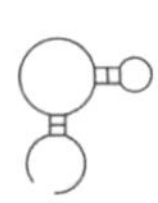	Shiratori I, *et al.* 2014

Aptamer	Nature	Sequence	Randomized region	Target	Function	Structure	Ref
RHA1635	DNA	GGGGCCCACCCTCTCG CTGGCGGCTCTGTTCTG TTCTCGTCTCCTTGATT TCTGTGGGCCCC	30 nt	rHA	IVA detection (ELAA)		Shiratori I, *et al.* 2014

Table 3. Aptamers against human IAV proteins.

Two DNA aptamers, A21 and A22, were isolated against an HA peptide containing amino acid positions 91–261. A22 was the most efficient aptamer to inhibit viral infection *in vivo* and *in vitro* by blocking the cellular receptor from binding HA. Moreover, A22 showed high binding activity against different IAV strains (H3N2 and H2N2) and reduced virus burden by 90%–99% in mice [61]. Further studies using the whole virus demonstrated the ability of RNA aptamers to distinguish between related strains within the H3N2 subtype of influenza type A viruses [62]. The selected aptamer P30-10-16 was able to discriminate between A/Panama/2007/1999 and A/Aichi/2/1968 H3N2 subtypes and its binding affinity to HA was even 15-fold higher compared with a monoclonal antibody specific to HA. A consensus aptamer sequence (5′-GUCGNCNU$(N)_{23}$GUA-3′) was selected by surface plasmon resonance (SPR) using an RNA pool based on randomized P30-10-16 (doped RNA pool). The GNCNU sequence was identified as the minimal element required to bind HA [63], suggesting a potential use as tools for influenza virus genotyping.

An aptamer selected against H5N1 HA (A10), showed inhibition of receptor binding producing *in vitro* inhibition of viral infection [64]. To increase the specificity, some aptamers were isolated using recombinant HA in the initial selection cycles and then the whole inactivated H5N1 virus for further selection cycles. The selected aptamers were able to discriminate among H5N2, H5N3, H5N9, H9N2 and H7N2, showing better specificity than anti-H5N1 monoclonal antibodies [65]. These aptamers were used on quartz crystal microbalance (QCM) biosensors coated with hydrogel. The hydrogel consisted of cross-linked hybridized ssDNA and aptamer. In the presence of H5N1 the hybridization is disturbed producing hydrogel swelling which is detected by a QCM sensor [66].

Although some aptamers have been isolated to bind a specific IVA subtype, some others identify more than one virus subtype. A 113-nt-long RNA aptamer (8-3) was isolated against HAs from H5-N1 and H7N7. The full 8-3 and shortened version called 8-3S aptamer were able to bind HA with high affinity and interfere with the cell surface HA–glycan interaction, suggesting a potential application in diagnosis and interference of virus–host interactions [67]. Furthermore, DNA aptamers were selected against recombinant hemagglutinin (rHa) to detect different subtypes of IVA such as H5N1, H1N1 and H3N2. The selected DNA aptamers: RHA0006, RH0385 and RHA1635 were able to successfully bind the three mentioned IVA subtypes. RHA0006 and RH0385 were also used in a sandwich enzyme-linked aptamer assay (ELAA), developing a novel, rapid and cost-effective diagnostic tool to identify various IVA subtypes [68].

2.3.2. *Influenza B Virus (IVB)*

Some aptamers have been isolated against whole virus or purified proteins in order to discriminate IVB from IVA (Table 4). An RNA aptamer against HA B/Johannesburg/05/1999 virus was able to discriminate between the HA from different strains and prevented viral infection by membrane fusion inhibition [62]. Two aptamers have been selected against intact HA of influenza strains B/Tokyo/S3/99 and Jilin/20/2003. The sensitivity of Tokyo aptamer was approximately 250-fold higher than a commercial antibody, demonstrating its potential to detect influenza viruses [69].

Aptamer	Nature	Sequence	Randomized region	Target	Function	Structure	Ref
Class A-20	RNA	GGGAGCUCAGCCUUCAC UGCACUCCGGCUGGUGG ACGCGGUACGAGCAAUU UGUACCGGAUGGAUGU UCGGGCAGCGGUGUGGC AGGGAUGAGCGGCACCA CGGUCGGAUCCAC	74 nt	HA	Discriminate between stran A and B. Membrane fusion inhibition	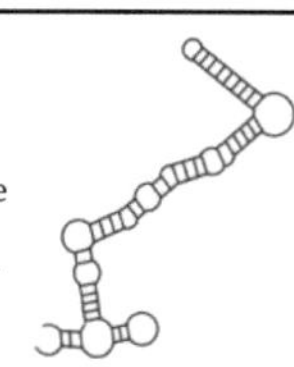	Gopinath JC, *et al.* 2006
Tokio virus aptamer (clone D)	RNA	GGGAGAAUUCCGACCAG AAGUUUUUGUUUAUAU UGUUGUUUUAUUCCUU UCCUCUCCUUCCUCUUC U	25 nt	Whole virus (Tokio virus)	Dscriminaton of influenza viruses and detection	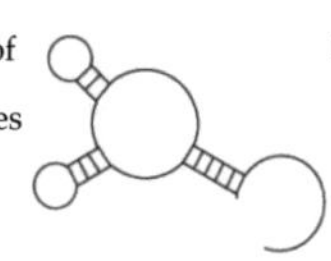	Lakshmipriya T, *et al.* 2013
Jilin-HA aptamer	RNA	GGGAGAAUUCCGACCAG AAGGGUCUACGCCCGAA GGGUUGCCGUGCCUUUC CUCUCUCCUUCCUCUUC U	25 nt	HA (Jilin HA)	Dscriminaton of influenza viruses and detection	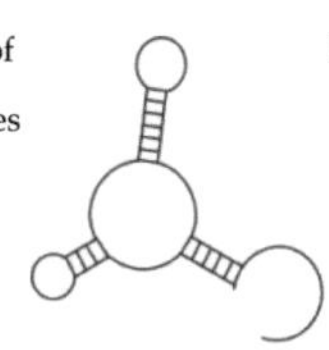	Lakshmipriya T, *et al.* 2013

Table 4. Aptamers against IBV.

2.4. Aptamers against HCV

HCV is one of the causes of chronic liver disease associated with end-stage cirrhosis and hepatocellular carcinoma. HCV are small enveloped viruses with a linear single-stranded RNA + genome containing a single ORF encoding a polyprotein flanked by untranslated regions (UTR) and processed into three structural proteins (C, E1 and E2) and seven nonstructural proteins (p7, NS2, NS3, NS4A, NS4B, NS5A and NS5B). The 5′-UTR contains an internal ribosomal entry site (IRES) important for mediated translation by association with the host cell small ribosomal unit (40S). Due to its importance in viral infection, replication and proliferation, HCV aptamers have been mainly isolated against NS3, NS5 proteins and some IRES domains (Table 5).

Aptamer	Nature	Sequence	Randomized region	Target	Function	Structure	Ref
10-G1	RNA	GGGAACUCGAUGAAGCGA AUUCUGUUGGCGAACUGU ACGCAAGUACACUGGAUG ACAGCCUAUCUAUCUAUC GGAUCCACG	10-18 nt	NS3	Inhibition of in vitro activity		Urvil PT, *et al.* 1997
G6-16	RNA	GGGAGAAUUCCGACCAGA AGGCUUGCUGUUGUUUCC CUGUUGUUUUGUCUCUCA ACUUUAUUGUGGUAAAGA UCACUGGGUUGAUAAGGG CUAACUCUAAUUUGACUA CAUGGUCGGACCAAUCAG UUCUUAUGGGAGAUGCAU AUGUGCGUCUACAUGGAU CCUCA	120 nt	NS3	Inhibition of proteolytic activity	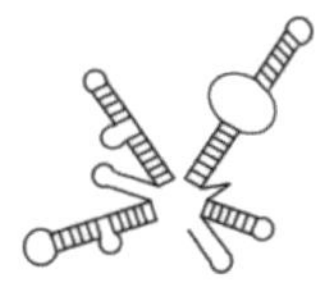	Kumar PK, *et al.* 1997
G6-19	RNA	GGGAGAAUUCCGACCAGA AGCUCUUAUACUAUUAAC GCUACCGUGUCAUUGUAC UUGGUAGUGUUGAUGGUU UGGGUCGCAUUUGGCUUG GCUUAUGGUUUUUUCACC CUACCUCUCAUUGACGCA GUAGGCUCUCAUAUGUGC GUCUACAUGGAUCCUCA	120 nt	NS3	Inhibition of proteolytic activity	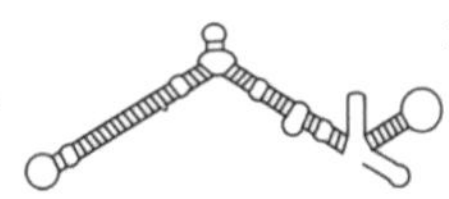	Kumar PK, *et al.* 1997
G9-I	RNA	GGGAGAAUUCCGACCAGA AGCUUCGGGAUUUGAGGG UAGAAUGGGACUACCUUU CCUCUCUCCUUCCUCUUC U	30 nt	ΔNS3	Inhibition of proteolytic activity	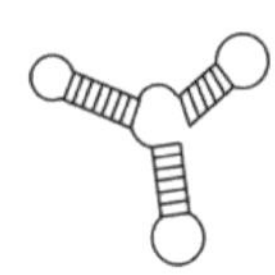	Fukuda K, *et al.* 2000
G9-II	RNA	GGGAGAAUUCCGACCAGA AGUGCUCUUAGAAUGGGA CUAAGACACGGGACCCUU UCCUCUCUCCUUCCUCUU CU	30 nt	ΔNS3	Inhibition of proteolytic activity	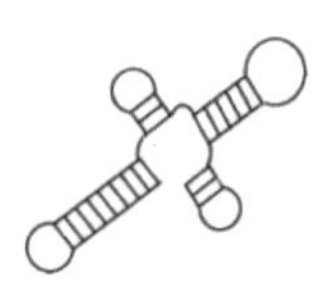	Fukuda K, *et al.* 2000
G9-III	RNA	GGGAGAAUUCCGACCAGA AGUACGACACGAUUGGGA CGUGUCUAUGGGACCCUU UCCUCUCUCCUUCCUCUU CU	30 nt	ΔNS3	Inhibition of proteolytic activity	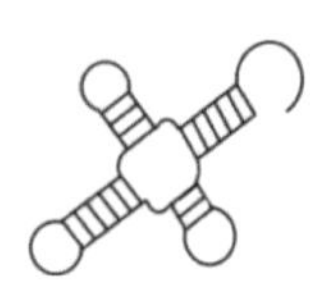	Fukuda K, *et al.* 2000

http://dx.doi.org/10.5772/61773

Aptamer	Nature	Sequence	Randomized region	Target	Function	Structure	Ref
B-2	RNA	GGGAUGCUUCGGCAUCCC CGAAGCCGCUAUGGACCA GUGGCGCGGCUUCGGCCC GACGGAGUGGUACCGCUU CGGCGGUACGUAAGCUUG GG	25 nt - 10 nt	NS5B	Inhibition of RNA polimerase activity in vitro		Biroccio A, *et al.* 2000
r10/43	DNA	GGGAGACAAGAATAAACG CTCAAGGGCGTGGTGGGTG GGGTACTAATAATGTGCGT TTGTTCGACAGGAGGCTCA CAACAGGC	36 nt	NS5B	Inhibition of specific subtype 3a polymerase activity	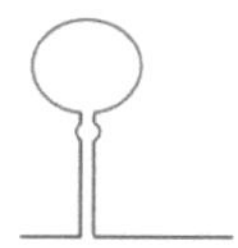	Jones LA, *et al.* 2006
r10/47	DNA	GGGAGACAAGAATAAACG CTCAATTGGGGTCTGCTCG GGATTGCGGAGAACGTGA ATCTTTCGACAGGAGGCTC ACAACAGGC	36 nt	NS5B	Inhibition of specific subtype 3a polymerase activity	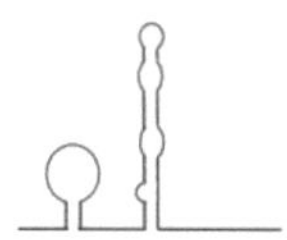	Jones LA, *et al.* 2006
NS2-2	DNA	CAGGTACCACCTTCATGGG CGCGGAAGACGATGGTGTA CTA	40 nt	NS2	Distrup Ns2 - Ns5b interaction. Inhibition of NS2 activity	ND	Gao Y, *et al.* 2014
NS2-3	DNA	ACGGGGCAGGATTGTCCCC GCGCCTGGTTGAAGGTAGT CGC	40 nt	NS2	Inhibithion of NS2 activity	ND	Gao Y, *et al.* 2014
ZE2	DNA	GCGGAATTCTAATACGACT CACTATAGGGAACAGTCCG AGCCGAATGAGGAATAATC TAGCTCCTTCGCTGAGGGT CAATGCGTCATAGGATCCC GC	30 nt	E2	Competitive inhibithion of E2 - CD81 binding . Block HCV infection.	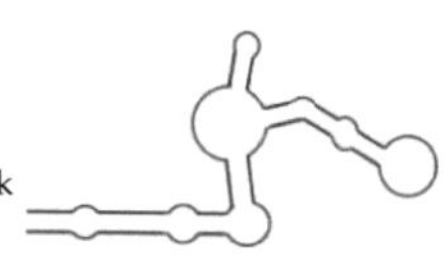	Chen F, *et al.* 2009
E1E2-6	DNA	ACGCTCGGATGCCACTACA G(N40)CTCATGGACGTGCTG GTGAC	40 nt	E1E2	Inhibition of aptamer binding to the host cell	ND	Yang D, *et al.* 2013

Table 5. Aptamers isolated against HCV proteins.

NS3 has a trypsin-like serine protease and NTPase/helicase activity [70, 71]. NS3 is required for proteolytic processing of nonstructural proteins [72]. The HCV protease domain disrupts the interferon (IFN) and toll-like receptor-3 (TLR3) signaling pathways by cleaving the caspase recruitment domain of mitochondrial antiviral signaling protein (MAVS) and the TIR domain containing an adapter-inducing interferon-β sequence (TRIF) [73]. As NS3 activity is crucial for viral replication, many aptamers have been isolated against NS3.

The 10G-1 RNA aptamer was selected against NS3 protease domain using a 12–18 nt randomized library and can reduce protease activity by 20% compared with serine protease inhibitors [74]. However, a new SELEX protocol was used to select anti-NS3 aptamers with improved binding and inhibition activities increasing the structural pool complexity by using larger randomized domains (120 nt) and competition against 10G-1. As a result, two new RNA aptamers were selected (G6–16 and G6–19) showing efficient NS3 binding. The G6–16 concentration needed to inhibit 50% of the NS3 activity was 3 μM, and although both aptamers inhibited the protease and helicase activity, they showed lower efficacy compared with known serine protease inhibitors [75]. To further improve aptamer efficacy and inhibit the NS3 RNA binding helicase, a truncated form (ΔNS3) only including the protease domain was used as a target, and the random sequence of the RNA pool was reduced to 30 nt to ease the SELEX process. Three highly specific aptamers (G9-I, G9-II and G9-III) were obtained against ΔNS3 containing the conserved sequence GA(A/U)UGGGAC that was present inside an identical loop in all aptamer structures. These aptamers showed K_D values of 11.6 nM, 6.3 nM and 8.9 nm, respectively. The G9 aptamers produced 90% of NS3 protease activity inhibition alone and 70% in the presence of NS4A used to simulate physiological conditions [76]. Structure analyses suggested that interaction of stem I and stem-loop II is essential to G9-I aptamer NS3 binding. To achieve *in vivo* applications, the G9-II aptamer was conjugated with cis-acting genomic human hepatitis delta virus (HDV) ribozymes. The aptamer was inserted into the nonfunctional stem IV region of the HDV ribozyme promoting *in vivo* stable structure that lasted up to 4 days after transfection. The HDV ribozyme–G9-II aptamer (HA) was attached to nuclear export signal CTEM45 (HAC) and ligated in tandem to increase the aptamers dosage in cells. These new constructs showed efficient NS3 protease inhibition *in vivo* and *in vitro* [77]. To also inhibit NS3 helicase activity, a poly U tail (14U) was added to the minimum functional sequence of the G9-I aptamer (ΔNEOIII) to mask and inhibit the helicase substrate-binding region [78]. NEOIII-14U displayed dual functions by inhibiting NS3 protease activity *in vivo* and *in vitro* and inhibiting the NS3 unwinding helicase reaction (IC_{50} 1 μM) [79].

More RNA aptamers were selected against NS3 helicase domain, including the conserved sequence GGA(U/C)GGAGCC at stem-loop regions. Further deletion and mutagenesis analyses demonstrated that the whole structure of the conserved stem-loop is needed for helicase inhibition. Aptamer #5 presented the best inhibition of helicase *in vitro* activity with an IC_{50} of 50 nM [77]. Bifunctional aptamers constructed conjugating RNA aptamers ΔNEOIII and G9-II with aptamer #5 through an oligo U spacer. The spacer length was optimized by protease and helicase inhibition assays [80]. The resulting advanced dual-functional (ADD) aptamers (NEO-34-s41 and G925-s50) showed superior inhibitory activities of NS3 [81].

NS5B is an RNA-dependent RNA polymerase that synthesizes the HCV-negative strand RNA using genomic positive RNA strand as a template. NS5B has an essential role in the HCV's life cycle and its variability has been associated with worse disease prognosis [82]. The highly specific B.2 RNA aptamer selected against a truncated NS5B target (NS5BΔC55) presented a conserved sequence that was folded on stem loop structure associated with a tight interaction to NS5B ($K_D = 1.5 \pm 0.2$ nM). Also, B.2 demonstrated inhibition of NS5B activity by a noncompetitive mechanism [9].

Two DNA aptamers selected against NS5B (27v and 127v) showed inhibition of polymerase activity *in vitro*. Although both aptamers were isolated from the same SELEX procedure and presented an 11-nt conserved sequence, they displayed different mechanisms to inhibit NS5B. The 27v aptamer competed with RNA template and inhibited both initiation and elongation of RNA synthesis, while 127v competed poorly and just inhibited initiation. Also, 27v was able to inhibit RNA synthesis and HCV particles production on Huh7 cells [83, 84]. The RNA aptamers r10/43 and r10/47 were isolated against NSB5 of HVC subtype 3a and resulted in the inhibition of polymerase activity with an estimated K_D = 1.4 and 6.0 nM, respectively [85].

In a different approach, chemically modified RNA aptamers (2′-hydroxyl or 2′-fluoropyrimidine) were isolated against NS5B. The 2′-hydroxyl aptamer inhibited HCV replication on human liver cells without producing off-target effects or generation of escape mutants. The 2′-fluoropyrimidine aptamer showed increased affinity to NS5B and efficient inhibition of HCV replication in cultured cells. This last aptamer was further conjugated with cholesterol or galactose-polyethylene glycol ligand to increase its availability and specificity for the liver inhibiting replication of HCV genotype 1b and 2a [86].

Two other RNA aptamers targeting NS5A (NS5A-4 and NS5A-5) reduced the levels of intracellular infectious virions and viral RNAs by 3-fold and 1-fold, respectively, affecting virus assembly and release through prevention of the NS5A–core protein interaction. These NS5A aptamers were specific to HCV without affecting HBV replication and produced cytotoxicity in human hepatocytes [87].

NS2 contains a transmembrane segment in the N-terminal and a cytoplasmic region in the C-terminal domain. Although NS2 is essential for HCV RNA replication, its role in HCV's life cycle is still unknown. Aptamers NS2-2 and NS2-3 were isolated against NS2 and demonstrated reduced infectious virus production without *in vitro* cytotoxicity. These aptamers were specific to HCV and did not trigger innate immunity responses. NS2-2 aptamer produces its antiviral effects through binding the NS2 N-terminus thus disrupting NS2–NS5 interaction [88].

E2 is an enveloped glycoprotein implicated on initial steps of viral infection by the direct interaction with CD81. Through cell surface SELEX (CS-SELEX), specific DNA aptamers were isolated against E2 expressed on CT26 cells. Aptamer ZE2 showed the highest affinity and specificity to E2 and was able to detect HCV particles and block HCV infection on human cultured hepatocytes by CD81 binding inhibition [89]. A similar inhibition mechanism was observed on the DNA aptamer E1E2-6, which inhibited viral infection by blocking host cell binding [90]. A new system developed to quantify immobilized infectious HCV particles in microplates (so-called enzyme linked apto-sorbent assay or ELASA) used aptamers against E2 instead of antibodies and resulted in an effective and easy-to-use tool to quantify infectious units of HCV and to monitor anti-HCV drug efficacies [91].

3. Aptamer structures

Nucleic acid aptamers have a diverse range of secondary structures such as stems, loops, symmetric or asymmetric internal loops, bulge, single-base bulges and junctions. Aptamer internal loops and bulges generally present different conformations in solution and adopt defined secondary and tertiary structures on ligand–aptamer complex [92]. This effect was observed on aptamer Sc5-c3 selected against HPV-16 VLPs. Sc5-c3 showed a hairpin structure with an internal loop, where the main loop (ML) presented two different structures in the absence of a target (Table 2). Sc5-c3 transition structure was demonstrated by ribonuclease mapping. Further experiments using Sc5-c3 mutants generated both stable stem and stable loop conformations, demonstrating that the loop structure binds better to the VLPs [58]. Thus, as observed in several aptamers, the binding region remains as a flexible single strand as bulges or loops stabilize conformation arrangements in the presence of a target, producing a very specific binding.

Although bulges and loops are quite common target-binding motifs in aptamer RNAs, they are not the only structures present in aptamer–target complexes. Pseudoknots and G-quadruplexes have also been reported as functional components of aptamers [93]. For example, some of the aptamers isolated against HIV integrase (93 del and 112 del) presented a G-rich nucleic acid sequence that was stabilized in the presence of K+ as G-tetrad, increasing their inhibitory effect [25]. Later reports showed that 93 del adopts an unusually stable dimeric quadruplex structure [94].

The binding properties of an aptamer are dictated by its sequence and subsequent folding into secondary and tertiary structures. Recently, functional RNA structures were classified as critical, connecting, neutral and forbidden structures regarding their particular roles within a structure [95]. This classification is also applicable to nucleic acid aptamers and is an important clue to design novel and functional variants for viral detection or therapy.

4. Challenges for aptamer technology

According to their molecular characteristics, RNA or DNA aptamers have some limitations in their use in animal models and humans. They have limited stability in biological fluids and are readily degraded by nucleases, unmodified aptamers in the bloodstream possess a half-life time of less than two minutes. However, many post-SELEX modifications have been developed to avoid nuclease attack and improve stability in biological fluids. Some modification examples include nucleotide substitutions by 2′-modified variants such as 2′-fluoro (2′-F), 2′-amino (2′-NH2) or 2′-O-alkyl. Because the most abundant nucleases in biological fluids are specific to pyrimidines, substitutions in pyrimidine positions appear to be sufficient to prevent degradation. Another method to stabilize RNA aptamers is the substitution of D-ribose by L-ribose. As a first step, the aptamers bind the mirror image of the target molecule to obtain a D-aptamer, then the selected aptamer sequence is synthesized in L-conformation. As a result of molecular symmetry, the L-ribose–containing aptamer can bind to the target molecule avoiding degradation by D-ribose–specific nucleases. Moreover, to efficiently overcome binding issues produced by the introduction of modified nucleotides on the aptamer sequence, the SELEX procedure can be carried out in the presence of modified libraries.

http://dx.doi.org/10.5772/61773

Therapeutic aptamers selected against intracellular or nuclear proteins represent bigger challenges as they need to go across physiological barriers (i.e. cell membrane) before they reach their targets. DNA and RNA aptamers are characterized by rapid renal clearance leading to short half-lives in the bloodstream. To address this issue, aptamers can be conjugated to synthetic polymers such as polyethylene glycol (PEG) to increase their *in vivo* half-life and pharmacodynamics [96]. Additionally, PEG-conjugated aptamers show higher cellular uptake than the unconjugated form [96]. Alternatively, delivery systems such as viral and nonviral vectors may have improved aptamer cell uptake and nuclear distribution [97]. Vectors or aptamers alone can be delivered either *ex vivo* or *in vivo*. *In vivo* approaches include intravenous injection or local implantation and *ex vivo* refer to the removal of cells followed by *in vitro* genetic manipulation and the reintroduction of modified cells. These therapies are still under evaluation and further studies are necessary to demonstrate their clinical safety. Aptamers against extracellular or surface viral targets have obvious advantages over aptamers targeting viral proteins intracellularly expressed, as they can reach exposed areas of infection, such as the respiratory tract or reproductive organs. This availability makes it possible to develop new antiviral drugs administrated by noninvasive methods, such as aerosols in case of respiratory tract infection or topical creams/lotions in case of reproductive organ infections. Many aptamers are undergoing clinical trials, some of them administrated by noninvasive methods but, so far, no antiviral aptamer has been approved for human use [18, 98].

5. Conclusion

In the last few years, aptamers have become successful tools for specific viral diagnosis and genotyping, resulting in the development of many methods based on aptamer–target detection with very high sensibility and accuracy. On the other hand, aptamer's role as an antiviral drug has demonstrated the inhibition of viral infection through *in vitro* assays and *in vivo* experiments using cell lines or animal models. Nevertheless, most aptamers failed to produce results in clinical trials mostly due to nuclease-associated degradation. Therefore, further development of aptamer's stability in biofluids and improved pharmacodynamics and delivery methods are required to overcome clinical issues that would allow its successful therapeutic application.

Author details

Ana Gabriela Leija-Montoya[1], María Luisa Benítez-Hess[2] and Luis Marat Alvarez-Salas[2*]

*Address all correspondence to: lalvarez@cinvestav.mx

1 Facultad de Medicina, Universidad Autónoma de Baja California, Mexicali B.C., México

2 Laboratorio de Terapia Génica, Departamento de Genética y Biología Molecular, Centro de Investigación y de Estudios Avanzados del I.P.N., México D.F., México

References

[1] Ellington AD, Szostak JW. In vitro selection of RNA molecules that bind specific ligands. Nature. 1990;346:818–822.

[2] Tuerk C, Gold L. Systematic evolution of ligands by exponential enrichment: RNA ligands to bacteriophage T4 DNA polymerase. Science. 1990;249:505–510.

[3] Aquino-Jarquin G, Toscano-Garibay JD. RNA aptamer evolution: two decades of SELEction. Int. J. Mol. Sci. 2011;12:9155–9171.

[4] Fitzwater T, Polisky B. A SELEX primer. Methods Enzymol. 1996;267:275–301.

[5] Bock LC, Griffin LC, Latham JA, Vermaas EH, Toole JJ. Selection of single-stranded DNA molecules that bind and inhibit human thrombin. Nature. 1992;355:564–566.

[6] Harada K, Frankel AD. Identification of two novel arginine binding DNAs. EMBO J. 1995;14:5798–5811.

[7] Wang Y, Killian J, Hamasaki K, Rando RR. RNA molecules that specifically and stoichiometrically bind aminoglycoside antibiotics with high affinities. Biochemistry. 1996;35:12338–12346.

[8] Sassanfar M, Szostak JW. An RNA motif that binds ATP. Nature. 1993;364:550–553.

[9] Biroccio A, Hamm J, Incitti I, De Francesco R, Tomei L. Selection of RNA aptamers that are specific and high-affinity ligands of the hepatitis C virus RNA-dependent RNA polymerase. J. Virol. 2002;76:3688–3696.

[10] Green LS, Jellinek D, Bell C, Beebe LA, Feistner BD, Gill SC, Jucker FM, Janjic N. Nuclease-resistant nucleic acid ligands to vascular permeability factor/vascular endothelial growth factor. Chem. Biol. 1995;2:683–695.

[11] Jellinek D, Green LS, Bell C, Lynott CK, Gill N, Vargeese C, Kirschenheuter G, McGee DP, Abesinghe P, Pieken WA. Potent 2′-amino-2′-deoxypyrimidine RNA inhibitors of basic fibroblast growth factor. Biochemistry. 1995;34:11363–11372.

[12] Bruno JG, Kiel JL. In vitro selection of DNA aptamers to anthrax spores with electrochemiluminescence detection. Biosens. Bioelectron. 1999;14:457–464.

[13] Ulrich H, Magdesian MH, Alves MJ, Colli W. In vitro selection of RNA aptamers that bind to cell adhesion receptors of *Trypanosoma cruzi* and inhibit cell invasion. J. Biol. Chem. 2002;277:20756–20762.

[14] Sundaram P, Kurniawan H, Byrne ME, Wower J. Therapeutic RNA aptamers in clinical trials. Eur. J. Pharm. Sci. 2013;48:259–271.

[15] Vinores SA. Pegaptanib in the treatment of wet, age-related macular degeneration. Int. J. Nanomedicine. 2006;1:263–268.

[16] Shum KT, Zhou J, Rossi JJ. Aptamer-based therapeutics: new approaches to combat human viral diseases. Pharmaceuticals (Basel). 2013;6:1507–1542.

[17] Pan W, Craven RC, Qiu Q, Wilson CB, Wills JW, Golovine S, Wang JF. Isolation of virus-neutralizing RNAs from a large pool of random sequences. Proc. Natl. Acad. Sci. U. S. A. 1995;92:11509–11513.

[18] Wandtke T, Wozniak J, Kopinski P. Aptamers in diagnostics and treatment of viral infections. Viruses. 2015;7:751–780.

[19] Barre-Sinoussi F, Chermann JC, Rey F, Nugeyre MT, Chamaret S, Gruest J, Dauguet C, Axler-Blin C, Vezinet-Brun F, Rouzioux C, Rozenbaum W, Montagnier L. Isolation of a T-lymphotropic retrovirus from a patient at risk for acquired immune deficiency syndrome (AIDS). Science. 1983;220:868–871.

[20] Gallo RC, Sarin PS, Gelmann EP, Robert-Guroff M, Richardson E, Kalyanaraman VS, Mann D, Sidhu GD, Stahl RE, Zolla-Pazner S, Leibowitch J, Popovic M. Isolation of human T-cell leukemia virus in acquired immune deficiency syndrome (AIDS). Science. 1983;220:865–867.

[21] Gopinath SC. Antiviral aptamers. Arch. Virol. 2007;152:2137–2157.

[22] Whitcomb JM, Hughes SH. Retroviral reverse transcription and integration: progress and problems. Annu. Rev. Cell Biol. 1992;8:275–306.

[23] Allen P, Worland S, Gold L. Isolation of high-affinity RNA ligands to HIV-1 integrase from a random pool. Virology. 1995;209:327–336.

[24] Phan AT, Modi YS, Patel DJ. Propeller-type parallel-stranded G-quadruplexes in the human c-myc promoter. J. Am. Chem. Soc. 2004;126:8710–8716.

[25] De Soultrait VR, Lozach PY, Altmeyer R, Tarrago-Litvak L, Litvak S, Andreola ML. DNA aptamers derived from HIV-1 RNase H inhibitors are strong anti-integrase agents. J. Mol. Biol. 2002;324:195–203.

[26] Chou SH, Chin KH, Wang AH. DNA aptamers as potential anti-HIV agents. Trends. Biochem. Sci. 2005;30:231–234.

[27] Berkhout B, Silverman RH, Jeang KT. Tat trans-activates the human immunodeficiency virus through a nascent RNA target. Cell. 1989;59:273–282.

[28] Yamamoto R, Toyoda S, Viljanen P, Machida K, Nishikawa S, Murakami K, Taira K, Kumar PK. In vitro selection of RNA aptamers that can bind specifically to Tat protein of HIV-1. Nucleic. Acids. Symp. Ser. 1995;145–146.

[29] Yamamoto R, Katahira M, Nishikawa S, Baba T, Taira K, Kumar PK. A novel RNA motif that binds efficiently and specifically to the Ttat protein of HIV and inhibits the trans-activation by Tat of transcription in vitro and in vivo. Genes Cells. 2000;5:371–388.

[30] Yamamoto R, Baba T, Kumar PK. Molecular beacon aptamer fluoresces in the presence of Tat protein of HIV-1. Genes Cells. 2000;5:389–396.

[31] Tombelli S, Minunni M, Luzi E, Mascini M. Aptamer-based biosensors for the detection of HIV-1 Tat protein. Bioelectrochemistry. 2005;67:135–141.

[32] Rahim RA, Tanabe K, Ibori S, Wang X, Kawarada H. Effects of diamond-FET-based RNA aptamer sensing for detection of real sample of HIV-1 Tat protein. Biosens. Bioelectron. 2013;40:277–282.

[33] Tuerk C, MacDougal-Waugh S. In vitro evolution of functional nucleic acids: high-affinity RNA ligands of HIV-1 proteins. Gene. 1993;137:33–39.

[34] Jensen KB, Green L, MacDougal-Waugh S, Tuerk C. Characterization of an in vitro-selected RNA ligand to the HIV-1 Rev protein. J. Mol. Biol. 1994;235:237–247.

[35] Giver L, Bartel DP, Zapp ML, Green MR, Ellington AD. Selection and design of high-affinity RNA ligands for HIV-1 Rev. Gene. 1993;137:19–24.

[36] Symensma TL, Giver L, Zapp M, Takle GB, Ellington AD. RNA aptamers selected to bind human immunodeficiency virus type 1 Rev in vitro are Rev responsive in vivo. J. Virol. 1996;70:179–187.

[37] Jensen KB, Atkinson BL, Willis MC, Koch TH, Gold L. Using in vitro selection to direct the covalent attachment of human immunodeficiency virus type 1 Rev protein to high-affinity RNA ligands. Proc. Natl. Acad. Sci. U. S. A. 1995;92:12220–12224.

[38] Lee SW, Gallardo HF, Gilboa E, Smith C. Inhibition of human immunodeficiency virus type 1 in human T cells by a potent Rev response element decoy consisting of the 13-nucleotide minimal Rev-binding domain. J. Virol. 1994;68:8254–8264.

[39] Yamada O, Kraus G, Luznik L, Yu M, Wong-Staal F. A chimeric human immunodeficiency virus type 1 (HIV-1) minimal Rev response element-ribozyme molecule exhibits dual antiviral function and inhibits cell-cell transmission of HIV-1. J. Virol. 1996;70:1596–1601.

[40] Gervaix A, Li X, Kraus G, Wong-Staal F. Multigene antiviral vectors inhibit diverse human immunodeficiency virus type 1 clades. J. Virol. 1997;71:3048–3053.

[41] Konopka K, Lee NS, Rossi J, Duzgunes N. Rev-binding aptamer and CMV promoter act as decoys to inhibit HIV replication. Gene. 2000;255:235–244.

[42] Khati M, Schuman M, Ibrahim J, Sattentau Q, Gordon S, James W. Neutralization of infectivity of diverse R5 clinical isolates of human immunodeficiency virus type 1 by gp120-binding 2′F-RNA aptamers. J. Virol. 2003;77:12692–12698.

[43] Dey AK, Khati M, Tang M, Wyatt R, Lea SM, James W. An aptamer that neutralizes R5 strains of human immunodeficiency virus type 1 blocks gp120-CCR5 interaction. J. Virol. 2005;79:13806–13810.

[44] Cohen C, Forzan M, Sproat B, Pantophlet R, McGowan I, Burton D, James W. An aptamer that neutralizes R5 strains of HIV-1 binds to core residues of gp120 in the CCR5 binding site. Virology. 2008;381:46–54.

[45] Mufhandu HT, Gray ES, Madiga MC, Tumba N, Alexandre KB, Khoza T, Wibmer CK, Moore PL, Morris L, Khati M. UCLA1, a synthetic derivative of a gp120 RNA aptamer, inhibits entry of human immunodeficiency virus type 1 subtype C. J. Virol. 2012;86:4989–4999.

[46] Zhou J, Swiderski P, Li H, Zhang J, Neff CP, Akkina R, Rossi JJ. Selection, characterization and application of new RNA HIV gp 120 aptamers for facile delivery of Dicer substrate siRNAs into HIV infected cells. Nucleic Acids Res. 2009;37:3094–3109.

[47] Neff CP, Zhou J, Remling L, Kuruvilla J, Zhang J, Li H, Smith DD, Swiderski P, Rossi JJ, Akkina R. An aptamer-siRNA chimera suppresses HIV-1 viral loads and protects from helper CD4(+) T cell decline in humanized mice. Sci. Transl. Med. 2011;3:1–10.

[48] Zhou J, Neff CP, Swiderski P, Li H, Smith DD, Aboellail T, Remling-Mulder L, Akkina R, Rossi JJ. Functional in vivo delivery of multiplexed anti-HIV-1 siRNAs via a chemically synthesized aptamer with a sticky bridge. Mol. Ther. 2013;21:192–200.

[49] DiPaolo JA, Popescu NC, Alvarez-Salas LM, Woodworth CD. Cellular and molecular alterations in human epithelial cells transformed by recombinant human papillomavirus DNA. Crit. Rev. Oncog. 1993;4:337–360.

[50] Hildesheim A, Herrero R, Wacholder S, Rodriguez AC, Solomon D, Bratti MC, Schiller JT, Gonzalez P, Dubin G, Porras C, Jimenez SE, Lowy DR. Effect of human papillomavirus 16/18 L1 viruslike particle vaccine among young women with preexisting infection: a randomized trial. JAMA. 2007;298:743–753.

[51] Werness BA, Levine AJ, Howley PM. Association of human papillomavirus types 16 and 18 E6 proteins with p53. Science. 1990;248:76–79.

[52] Dyson N, Howley PM, Munger K, Harlow E. The human papilloma virus-16 E7 oncoprotein is able to bind to the retinoblastoma gene product. Science. 1989;243:934–937.

[53] Belyaeva TA, Nicol C, Cesur O, Trave G, Blair GE, Stonehouse NJ. An RNA aptamer targets the PDZ-binding motif of the HPV16 E6 oncoprotein. Cancers (Basel). 2014;6:1553–1569.

[54] Nicol C, Bunka DH, Blair GE, Stonehouse NJ. Effects of single nucleotide changes on the binding and activity of RNA aptamers to human papillomavirus 16 E7 oncoprotein. Biochem. Biophys. Res. Commun. 2011;405:417–421.

[55] Nicol C, Cesur O, Forrest S, Belyaeva TA, Bunka DH, Blair GE, Stonehouse NJ. An RNA aptamer provides a novel approach for the induction of apoptosis by targeting the HPV16 E7 oncoprotein. PLoS ONE. 2013;8:e64781–e64791.

[56] Toscano-Garibay JD, Benitez-Hess ML, Alvarez-Salas LM. Isolation and characterization of an RNA aptamer for the HPV-16 E7 oncoprotein. Arch. Med. Res. 2011;42:88–96.

[57] Toscano-Garibay JD, Benitez-Hess ML, Alvarez-Salas LM. Targeting of the HPV-16 E7 protein by RNA aptamers. Methods Mol. Biol. 2015;1249:221–239.

[58] Leija-Montoya AG, Benitez-Hess ML, Toscano-Garibay JD, Alvarez-Salas LM. Characterization of an RNA aptamer against HPV-16 L1 virus-like particles. Nucleic Acid Ther. 2014;24:344–355.

[59] Gourronc FA, Rockey WM, Thiel WH, Giangrande PH, Klingelhutz AJ. Identification of RNA aptamers that internalize into HPV-16 E6/E7 transformed tonsillar epithelial cells. Virology. 2013;446:325–333.

[60] Graham JC, Zarbl H. Use of cell-SELEX to generate DNA aptamers as molecular probes of HPV-associated cervical cancer cells. PLoS ONE. 2012;7:e36103–e36111.

[61] Jeon SH, Kayhan B, Ben-Yedidia T, Arnon R. A DNA aptamer prevents influenza infection by blocking the receptor binding region of the viral hemagglutinin. J. Biol. Chem. 2004;279:48410–48419.

[62] Gopinath SC, Misono TS, Kawasaki K, Mizuno T, Imai M, Odagiri T, Kumar PK. An RNA aptamer that distinguishes between closely related human influenza viruses and inhibits haemagglutinin-mediated membrane fusion. J. Gen. Virol. 2006;87:479–487.

[63] Misono TS, Kumar PK. Selection of RNA aptamers against human influenza virus hemagglutinin using surface plasmon resonance. Anal. Biochem. 2005;342:312–317.

[64] Cheng C, Dong J, Yao L, Chen A, Jia R, Huan L, Guo J, Shu Y, Zhang Z. Potent inhibition of human influenza H5N1 virus by oligonucleotides derived by SELEX. Biochem. Biophys. Res. Commun. 2008;366:670–674.

[65] Wang R, Zhao J, Jiang T, Kwon YM, Lu H, Jiao P, Liao M, Li Y. Selection and characterization of DNA aptamers for use in detection of avian influenza virus H5N1. J. Virol. Methods. 2013;189:362–369.

[66] Wang R, Li Y. Hydrogel based QCM aptasensor for detection of avian influenza virus. Biosens. Bioelectron. 2013;42:148–155.

[67] Suenaga E, Kumar PK. An aptamer that binds efficiently to the hemagglutinins of highly pathogenic avian influenza viruses (H5N1 and H7N7) and inhibits hemagglutinin-glycan interactions. Acta Biomater. 2014;10:1314–1323.

[68] Shiratori I, Akitomi J, Boltz DA, Horii K, Furuichi M, Waga I. Selection of DNA aptamers that bind to influenza A viruses with high affinity and broad subtype specificity. Biochem. Biophys. Res. Commun. 2014;443:37–41.

[69] Lakshmipriya T, Fujimaki M, Gopinath SC, Awazu K. Generation of anti-influenza aptamers using the systematic evolution of ligands by exponential enrichment for sensing applications. Langmuir. 2013;29:15107–15115.

[70] Gallinari P, Brennan D, Nardi C, Brunetti M, Tomei L, Steinkuhler C, De FR. Multiple enzymatic activities associated with recombinant NS3 protein of hepatitis C virus. J. Virol. 1998;72:6758–6769.

[71] Stapleford KA, Lindenbach BD. Hepatitis C virus NS2 coordinates virus particle assembly through physical interactions with the E1–E2 glycoprotein and NS3–NS4A enzyme complexes. J. Virol. 2011;85:1706–1717.

[72] Failla C, Tomei L, De FR. Both NS3 and NS4A are required for proteolytic processing of hepatitis C virus nonstructural proteins. J. Virol. 1994;68:3753–3760.

[73] Preciado MV, Valva P, Escobar-Gutierrez A, Rahal P, Ruiz-Tovar K, Yamasaki L, Vazquez-Chacon C, Martinez-Guarneros A, Carpio-Pedroza JC, Fonseca-Coronado S, Cruz-Rivera M. Hepatitis C virus molecular evolution: transmission, disease progression and antiviral therapy. World J. Gastroenterol. 2014;20:15992–16013.

[74] Urvil PT, Kakiuchi N, Zhou DM, Shimotohno K, Kumar PK, Nishikawa S. Selection of RNA aptamers that bind specifically to the NS3 protease of hepatitis C virus. Eur. J. Biochem. 1997;248:130–138.

[75] Kumar PK, Machida K, Urvil PT, Kakiuchi N, Vishnuvardhan D, Shimotohno K, Taira K, Nishikawa S. Isolation of RNA aptamers specific to the NS3 protein of hepatitis C virus from a pool of completely random RNA. Virology. 1997;237:270–282.

[76] Fukuda K, Vishnuvardhan D, Sekiya S, Hwang J, Kakiuchi N, Taira K, Shimotohno K, Kumar PK, Nishikawa S. Isolation and characterization of RNA aptamers specific for the hepatitis C virus nonstructural protein 3 protease. Eur. J. Biochem. 2000;267:3685–3694.

[77] Nishikawa F, Kakiuchi N, Funaji K, Fukuda K, Sekiya S, Nishikawa S. Inhibition of HCV NS3 protease by RNA aptamers in cells. Nucleic Acids. Res. 2003;31:1935–1943.

[78] Sekiya S, Nishikawa F, Fukuda K, Nishikawa S. Structure/function analysis of an RNA aptamer for hepatitis C virus NS3 protease. J. Biochem. (Tokyo.). 2003;133:351–359.

[79] Fukuda K, Umehara T, Sekiya S, Kunio K, Hasegawa T, Nishikawa S. An RNA ligand inhibits hepatitis C virus NS3 protease and helicase activities. Biochem. Biophys. Res. Commun. 2004;325:670–675.

[80] Umehara T, Fukuda K, Nishikawa F, Sekiya S, Kohara M, Hasegawa T, Nishikawa S. Designing and analysis of a potent bi-functional aptamers that inhibit protease and helicase activities of HCV NS3. Nucleic Acids Symp. Ser. (Oxf). 2004;195–196.

[81] Umehara T, Fukuda K, Nishikawa F, Kohara M, Hasegawa T, Nishikawa S. Rational design of dual-functional aptamers that inhibit the protease and helicase activities of HCV NS3. J. Biochem. (Tokyo). 2005;137:339–347.

[82] Marascio N, Torti C, Liberto M, Foca A. Update on different aspects of HCV variability: focus on NS5B polymerase. BMC Infect. Dis. 2014;14:S1–14.

[83] Bellecave P, Andreola ML, Ventura M, Tarrago-Litvak L, Litvak S, Astier-Gin T. Selection of DNA aptamers that bind the RNA-dependent RNA polymerase of hepatitis C virus and inhibit viral RNA synthesis in vitro. Oligonucleotides. 2003;13:455–463.

[84] Bellecave P, Cazenave C, Rumi J, Staedel C, Cosnefroy O, Andreola ML, Ventura M, Tarrago-Litvak L, Astier-Gin T. Inhibition of hepatitis C virus (HCV) RNA polymerase by DNA aptamers: mechanism of inhibition of in vitro RNA synthesis and effect on HCV-infected cells. Antimicrob. Agents Chemother. 2008;52:2097–2110.

[85] Jones LA, Clancy LE, Rawlinson WD, White PA. High-affinity aptamers to subtype 3a hepatitis C virus polymerase display genotypic specificity. Antimicrob. Agents Chemother. 2006;50:3019–3027.

[86] Lee CH, Lee YJ, Kim JH, Lim JH, Kim JH, Han W, Lee SH, Noh GJ, Lee SW. Inhibition of hepatitis C virus (HCV) replication by specific RNA aptamers against HCV NS5B RNA replicase. J. Virol. 2013;87:7064–7074.

[87] Yu X, Gao Y, Xue B, Wang X, Yang D, Qin Y, Yu R, Liu N, Xu L, Fang X, Zhu H. Inhibition of hepatitis C virus infection by NS5A-specific aptamer. Antiviral Res. 2014;106:116–124.

[88] Gao Y, Yu X, Xue B, Zhou F, Wang X, Yang D, Liu N, Xu L, Fang X, Zhu H. Inhibition of hepatitis C virus infection by DNA aptamer against NS2 protein. PLoS ONE. 2014;9:e90333–e90343.

[89] Chen F, Hu Y, Li D, Chen H, Zhang XL. CS-SELEX generates high-affinity ssDNA aptamers as molecular probes for hepatitis C virus envelope glycoprotein E2. PLoS ONE. 2009;4:e8142.

[90] Yang D, Meng X, Yu Q, Xu L, Long Y, Liu B, Fang X, Zhu H. Inhibition of Hepatitis C virus infection by DNA aptamer against envelope protein. Antimicrob. Agents Chemother. 2013;57:4937–4944.

[91] Park JH, Jee MH, Kwon OS, Keum SJ, Jang SK. Infectivity of hepatitis C virus correlates with the amount of envelope protein E2: development of a new aptamer-based assay system suitable for measuring the infectious titer of HCV. Virology. 2013;439:13–22.

[92] Patel DJ. Structural analysis of nucleic acid aptamers. Curr. Opin. Chem. Biol. 1997;1:32–46.

http://dx.doi.org/10.5772/61773

[93] Zhang Y, Yu Z, Jiang F, Fu P, Shen J, Wu W, Li J. Two DNA aptamers against avian influenza H9N2 virus prevent viral infection in cells. PLoS ONE. 2015;10:e0123060.

[94] Phan AT, Kuryavyi V, Ma JB, Faure A, Andreola ML, Patel DJ. An interlocked dimeric parallel-stranded DNA quadruplex: a potent inhibitor of HIV-1 integrase. Proc. Natl. Acad. Sci. U. S. A. 2005;102:634–639.

[95] Kun A, Szathmary E. Fitness landscapes of functional RNAs. Life (Basel). 2015;5:1497–1517.

[96] Da PC, Blackshaw E, Missailidis S, Perkins AC. PEGylation and biodistribution of an anti-MUC1 aptamer in MCF-7 tumor-bearing mice. Bioconjug. Chem. 2012;23:1377–1381.

[97] Silva AC, Lopes CM, Sousa Lobo JM, Amaral MH. Nucleic acids delivery systems: a challenge for pharmaceutical technologists. Curr. Drug Metab. 2015;16:3–16.

[98] Gopinath SC. Methods developed for SELEX. Anal. Bioanal. Chem. 2007;387:171–182.

Chapter 7

DNA G-Quadruplex as a Reporter System for Sensor Development

Qiuting Loh and Theam Soon Lim

Additional information is available at the end of the chapter

http://dx.doi.org/10.5772/61580

Abstract

The versatile DNA G-quadruplex structure has emerged as an interesting alternative reporter system applied in different biosensor platforms. In comparison to the conventional reporter systems like enzymatic or fluorescent, DNA G-quadruplex has some distinct advantages, as it is thermostable, easy to produce, low cost and most importantly able to be amplified. Such remarkable advantages have led many researchers to exploit DNA G-quadruplex as the reporter system in colorimetric, fluorescence and luminescence sensors. There has also been integration of DNA G-quadruplex with electrochemical methods and quantum dot for sensing applications. Therefore, this chapter highlights some recent examples of different biosensor platforms that use DNA G-quadruplex as a reporter system with different detection methods.

Keywords: G-quadruplex, Biosensor, Protein Detection, DNA Detection, Metal Detection

1. Introduction

All genetic information with regards to every living organism is stored in the deoxyribonucleic acid, DNA. This is the fundamental application of DNA that makes it the basic building block of life. However, DNA is a very dynamic molecule whereby its function is not confined only to information storage and delivery. DNA also has the ability to form a number of spatial arrangements such as single-stranded hairpins, homoduplexes, triplexes and quadruplexes with high-order complexity. In nature, the formation of these structures has been found to be involved in many cellular mechanisms such as DNA recombination, regulation of gene expression and possibly the proliferation of tumor cells [1, 2].

One of these DNA structures that are well studied is the G-quadruplex (G-quad) structure. The G-quad structure is made up of a stack of nucleic acid sequence that is rich in guanine (G)

[3]. Besides, G-quad structure is found to be polymorphic as it forms many different structural arrangements depending on the variation in DNA composition and environment. Different formation of G-quad structure has different yet specific functions in nature. Thus, the study of G-quad has driven the structure into many applications, especially in the field of medicine, biology and material sciences.

The biological functions of the G-quad structure are well documented and the principles of it can be exploited for use as biosensors and therapeutics. One of the functions of the structure is the formation of DNA enzyme or catalytic enzymes, in short, DNAzyme, that can exhibit catalytic capabilities such as exhibiting peroxidase-like activity. G-quad can also serve as internal fluorescent probes in which its nucleobases are modified or attached with fluorescent dyes for sensing. In addition, G-quad can be combined with some electrochemical methods to produce signal readout. The discovery of such remarkable advantages of G-quad have given rise to the development of assays exploiting the G-quad structure such as DNA detection assay, protein detection assay and even the detection of molecules and ions [4-7].

Taken together, the advantages and flexibilities accorded to DNA G-quad have made G-quad very useful for the development of a variety of reporter systems for sensing applications. Consequently, DNA-based assay has now become a potential alternative to the conventional diagnostic platforms that use enzymes. In this chapter, the focus of our discussion would be on the structural features and application of DNA G-quad structures for the development of various sensing platforms.

2. Basic structures of DNA G-quad

DNA is often described as a double helix structure based on the typical Watson-Crick base pairing, where hydrogen bonds are formed between guanine and cytosine or adenine with thymine [8]. However, it was later discovered that a different bonding interaction based on hydrogen bonding could contribute to base paring, called the Hoogsteen bonding. The basic structure of G-quad involves four G bases forming a square planar array called the G-quartet that is stabilized by eight Hoogsteen and Watson-Crick hydrogen bonds. G-quad are formed by stacking up square planar arrays that is joined by the phosphodiester backbone and stabilized by the π–π stacking interactions of the stacked G-quartets and specific cations such as K^+, Na^+, Li^+, NH_4^+, $Pb^{2+}$and Sr^{2+} that gives rise to the strong electrostatic interaction between G and the cations [9-10].

NMR and crystallography studies have shown G-quad structures to be highly polymorphic [11]. They can form many different structures depending on the length of the DNA, orientation of the chains, positions of the loops and nature of the cations. G-quad can be found in many different forms ranging from one, two or four separate chains that give rise to unimolecular, bimolecular and tetramolecular structures [12]. There are also a variety of topologies of G-quad due to different possible combinations of the stretches of G-rich sequences, loop formation and also sequences [13-14]. Generally, the stretches of G can fold into different forms that include the parallel, basket, hairpin and chair conformation. Four strands of G can fold to form a

http://dx.doi.org/10.5772/61580

parallel four-stranded structure while two chains of G will fold into dimeric structures by dimerization of a pair of hairpin that results in a bimolecular G-quad with two loops [15]. Two structures of different loop orientations can be formed which are 'edgewise' loops that connect adjacent anti-parallel chains and the 'diagonal' loops that connect cross-over anti-parallel chains. As reported, the use of different cations in these two structures will form different loop conformations. The NMR solution of these two structures showed that with K^+ ions, the G-quad structure gave 'edgewise' loops, and, on the other hand, the G-quad structure in Na^+ ions produced 'diagonal' loops [16, 17]. A single-stranded G-rich sequence with four G repeats will form a unimolecular G-quad structure. This single stretch of G will fold and form an intramolecular G-quad with three loops in the presence of cations. Due to the steric hindrance and electrostatic repulsion caused by the loops, the orientation of the three loops is not entirely anti-parallel [13].

The polymorphism in the G-quad is the result of a balance between several stabilizing factors. The G-quad structures are mainly determined by monovalent cations as these structures are cation-dependent and require it for stability. Besides, other factors such as hydrogen bonding, base-stacking forces and hydrophobic effects also affect the formation of different topologies of the G-quad structures. However, this remarkable polymorphism has driven the increasing influence of G-quad in various functions and applications, especially in the field of medicine, biology and material sciences [18]. The main attractive application of G-quad revolves around the potential diagnostic application of G-quad as a reporter system where signal readouts are easily amplified by standard DNA amplification processes to yield a sensitive sensing system.

3. G-quad mimicking peroxidase activity in colorimetric-based sensors

There are several forms of readouts that G-quad structures are capable of producing. A major readout format is by absorbance value where G-quad structures are able to form DNAzymes that exhibit catalytic activity [19]. In nature, DNAzymes are initially known as catalytic enzymes or DNA enzymes because of its ability to catalyze many reactions such as ligation, DNA modification [20, 21], cleavage of DNA or RNA [21, 22] and also methylation of porphyrin rings [9,23]. One of the most important features of G-quad DNAzyme activity is the peroxidase mimicking activity when hemin is bound to the G-quad structure. The hemin–quadruplex complex will catalyze the peroxide-mediated oxidation of the 2,2′-azino-bis(3-ethylbenzthiazoline-6-sulphonic acid) diammonium salt (ABTS) to generate a coloured product [24-26]. This colorimetric change has allowed the development of assays detecting metal ions, aptamer–substrate complexes and even proteins. The conventional method to generate such change in colour is based on enzymatic reactions involving enzymes such as horseradish peroxidase and alkaline phosphatase. These enzymes are usually active in a narrow temperature range and denatures at high temperatures. DNAzymes are stable in a broad temperature range and even at very high temperatures, making it an interesting alternative to enzymes. Another advantage of DNAzymes is the ease in preparation by chemical synthesis or by PCR, whereas protein enzymes require tedious preparation and purification processes. A key advantage of G-quad DNAzyme is the ability to carry out signal amplification by conventional DNA amplification methods, which is impossible with normal protein set-ups. The attractiveness of G-quad DNAzyme to mimic peroxidase-like activity has allowed it to be exploited for the development of biosensors [27-30].

An assay to detect silver ions (Ag^+) was developed using G-quad as the reporter system. Ag^+ ions are able to stabilize cytosine–cytosine (C–C) mismatches by forming the C–Ag^+–C base pairs. Therefore, in the absence of Ag^+, the G-rich sequence will form an intramolecular duplex. The addition of Ag^+ in the mixture will allow the G-rich sequence to fold into a quadruplex structure and readily bind to hemin to form a DNAzyme. This will allow the G-quad to exhibit the peroxidase-like activity resulting in a change of the colourless ABTS to a coloured product in the presence of Ag^+ ions. The application of G-quad is useful to detect other molecules other than metal ions [31].

Our group has shown the application of G-quad as a sensitive reporter system for the detection of antibody–antigen interaction. The system was based on a pre-formed reporter system whereby a probe was pre-formed by conjugating streptavidin gold nanoparticles with biotinylated antigen and biotinylated daunomycin aptamer to exhibit the hemin-dependent peroxidase-like activity (Figure 1). Thus, the pre-assay generation of such reporter probes allows for rapid one-step incubation in a one-pot synthesis by exploiting the simple yet strong streptavidin–biotin interaction. This helps to eliminate multiple tedious steps of incubation and wash of conventional immunoassay systems. In this direct antibody–antigen assay, antibodies against the target antigen was coated to the microtiter plates and incubated with the probe. The wells were then developed with ABTS solution. The assay was able to generate sensitive readouts for both competitive and direct assays [32].

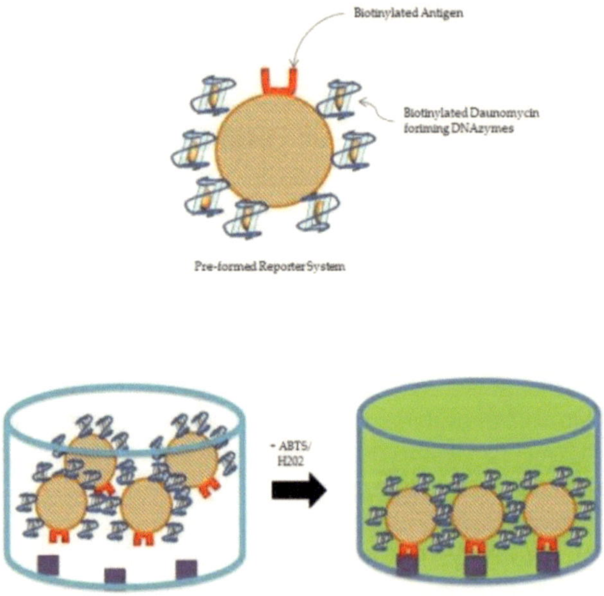

Figure 1. Schematic diagram of the immunoassay that is based on the preformed-reporter system.

Commonly, the peroxidase-like activity of DNAzymes is based on the oxidation of ABTS solution mediated by hydrogen peroxide (H_2O_2). This mechanism can be manipulated to initiate different sensing outcomes based on the same principles. This was evident in a detection assay for cholesterol by substituting the H_2O_2 with the catalysed cholesterol oxidase reaction of cholesterol and oxygen to produce the required H_2O_2 (Figure 2). Therefore, the G-quad DNAzyme will be able to exhibit the peroxidase-like activity after the binding of hemin to catalyze the oxidation of colourless ABTS mediated by the cholesterol oxidase reacted production of H_2O_2. Therefore, only with the presence of cholesterol, the changes in colour of ABTS will occur. Amplification of the DNA sequences to produce more DNAzyme was carried out to further improve the signal readout [33].

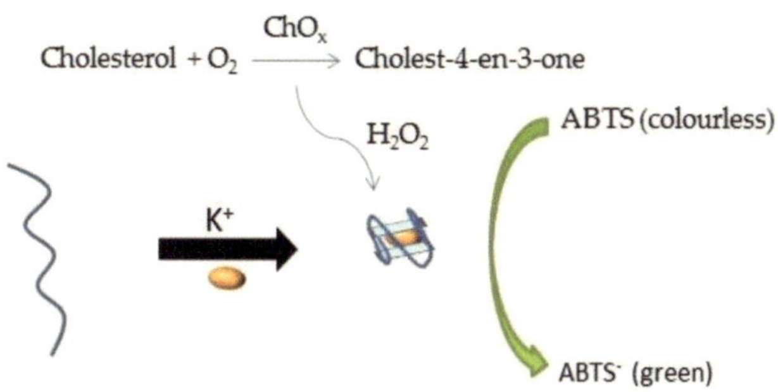

Figure 2. Schematic diagram of the G-quad-based colorimetric sensor for cholesterol.

A new strategy was reported by Tang et al. for nucleic acid detection of the viral load of Hepatitis B virus (HBV) that utilizes G-quad DNAzyme as the probe and involves polymerase chain reaction (PCR). The DNA probe was designed to form a hairpin structure at room temperature. During the denaturation step of PCR, the dsDNA HBV templates will denature, and the DNA probe will be linearized and allowed to hybridize with the target HBV ssDNA. The loop of the DNA probes will anneal to the conserved region of the HBV genome. The DNA probes can be amplified and cleaved so that they can form G-quad DNAzyme. The stem part of the probe is used to prohibit DNAzyme sequence to fold into G-quad at room temperature in order to reduce the background reading. With the involvement of PCR, the sensitivity and specificity of the HBV DNA assay was improved [34].

For further improvement of the DNA probe amplification, single temperature amplification (isothermal amplification) was carried out instead of going through the conventional temperature cycling with the PCR. As reported by Liu et al., an aptamer-modified microchip that combines rolling circle amplification with the peroxidase mimicking activity by G-quad DNAzyme was developed for thrombin detection (Figure 3). The assay involved an aptamer-modified microchip and a reporter aptamer consisting of the thrombin aptamer sequence and a primer with G-quad circular template. When the sample was introduced into the microchip, thrombin was captured by the chip immobilized thrombin aptamer and the reporter aptamer will act as primer for the RCA amplification process. This will allow the amplification process to generate increased amounts of G-quad to form DNAzyme upon binding of hemin. The generated hemin–DNAzyme complex will catalyse the oxidation of ABTS to produce a coloured product [35]. Isothermal DNA amplification strategies can be employed easily for DNAzyme sensor detection.

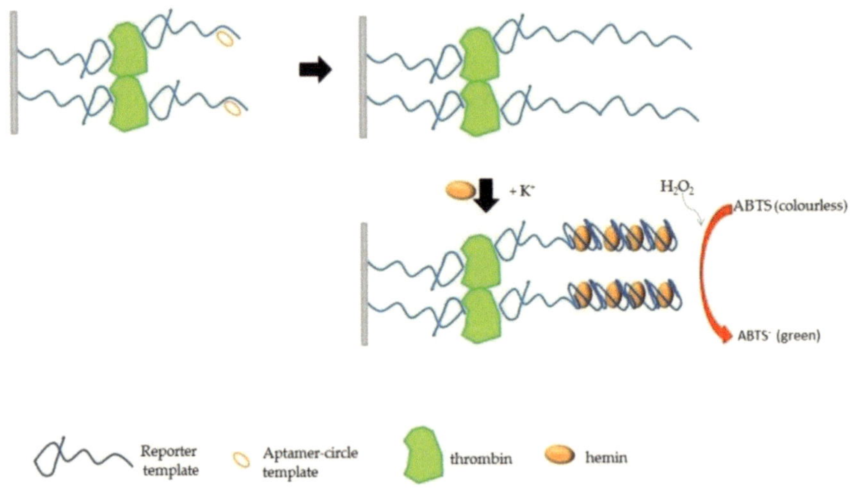

Figure 3. Ultrasensitive detection of thrombin by RCA and G-quad DNAzyme.

Besides RCA, G-quad structures can also be amplified using other isothermal strategies like quadruplex priming amplification (QPA) as reported by Kankia et al. QPA allows for efficient generation of G-quad structures by isothermal amplification with the additional ability to self-dissociate for continuous amplification. During the QPA elongation process, the 5′ end of the product will fold into intramolecular quadruplex and self-dissociates from the primer-binding site of the template, allowing the template to be accessible to the incoming primer for the next priming cycle. Thus, more quadruplexes were formed and eventually leading to increased signal readout. Our group incorporated the QPA system for the development of a sandwich immunoassay, called immuno-QPA (IQPA). The system exploits the peroxidase mimicking function of G-quad DNAzyme and the quadruplex amplification by QPA. A recombinant scFv was coated on the well and the biotinylated antigen was added and left to bind with the antibody. Streptavidin was introduced as a bridge between the biotinylated antigen and biotinylated QPA template. QPA was then carried out to amplify the QPA template with its specific primer to generate self-dissociating G-quad structures. These self-dissociating G-quad structures will bind to hemin to generate the similar colour change readout with ABTS as mentioned earlier [36].

4. G-quad as fluorescence probes

The photophysical properties of G-quad structures are highly dependent on the bound metal ions at the location of nucleobase electrons of the structure due to the coordination of ions in the center of the quadruplex. Thus, G-quad is able to exhibit two- to ten-fold higher quantum fluorescence yields. This is due to the structure serving as energy donors to energy acceptors in close proximity to yield fluorescence resonance energy transfer (FRET) systems. G-quad can be utilized as a fluorescent probe in which the nucleobases in the structure are labelled or attached with fluorescent dyes. Besides, the enhancement of the fluorescence of a small dye can be exhibited based on the interaction of the G-quad with the fluorescence ligand [7,37,38]. Some of the ligands such as thiazole orange were reported to yield higher fluorescence yield with the aid of the G-quad structure [39]. On the other hand, the G-quad could also cause fluorescence quenching phenomena. For instance, the classic intercalating agent, ethidium bromide (EtBr), binds to double-stranded DNA and exhibit fluorescence intensity up to 30-fold. However, in the presence of G-quad, the fluorescence intensity of the EtBr can be quenched upon binding to the structure [7].

One of the label-free G-quad DNA-based fluorescence biosensors reported was designed to detect cisplatin, which is an anticancer drug widely used in chemotherapy. However, cisplatin overdose can cause neurotic cell death that makes it dangerous for administration without proper monitoring. Therefore, Zhou et al. developed a simple label-free fluorescence biosensor by utilizing the interaction of G-quad and cisplatin as cisplatin was found to bind to G-quad naturally (Figure 4). In the context of the assay, the absence of cisplatin will allow the formed G-quad to bind to N-methyl mesoporphyrin IX (NMM), resulting in an increase in fluorescence intensity. However, when cisplatin was added into the assay, it will bind to the G-quad structure and disintegrate the structure. This will then cause a drastic decrease in fluorescence intensity due to the collapse of the G-quad structure [40].

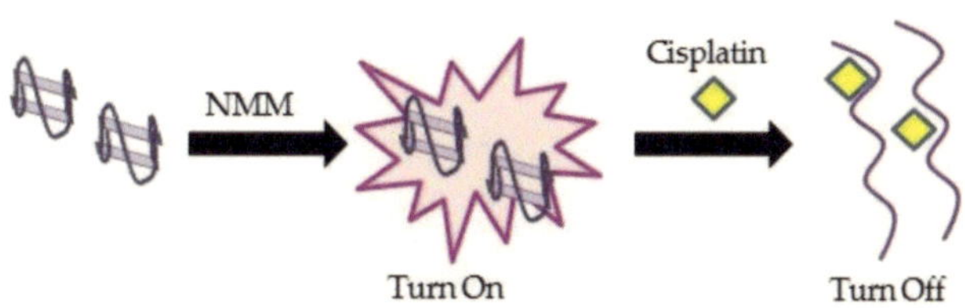

Figure 4. Label-free turn on and turn off fluorescence DNA G-quad-based sensor for the detection of cisplatin.

Wang et al. had proposed a label-free fluorescence biosensor based on G-quad formation in order to detect the lead (II) ion (Pb^{2+}). This biosensor consists of a G-rich DNA strand with its partially complementary strand. The heme-oxygenase-1 inhibitor, zinc protoporphyrin (ZnPPIX) has been utilized as a fluorescence probe, whereby it can interact with the Pb^{2+}/G–quad complex, producing fluorescence readout. In the absence of the Pb^{2+} ions, both DNA strands will form a DNA duplex. The addition of Pb^{2+} will unwind the DNA duplex to allow the G-rich strand to fold into the G-quad structure. The Pb^{2+}/G–quad complex will then interact with ZnPPIX to enhance its fluorescence intensity, which is not possible with the DNA duplex. This biosensor could overcome the cumbersome step in which most of the fluorescence-based biosensors would require fluorescence dyes to be probed on the DNA. In addition, this biosensor can be reset easily to the original state by dissociating the G-quad structure. In order to dissociate the G-quad structure, the strong Pb^{2+} chelator DOTA (1,4,7,10-tetraazacyclododecane-1,4,7,10-tetraacetic acid) was used as it has high binding capacity against Pb^{2+} to remove Pb^{2+} from the G-quad structure, leaving the G-quad structure to dissociate and form a DNA duplex with the complementary strand again [41].

As reported by Kankia and co-workers, a further improvement to their original QPA method was reported [42]. The improved method involves the use of two linear processes that are the QPA and linear nicking amplification, which is based on the study done by Galas and co-workers. The probe DNA was hybridized with the target DNA and polymerase was used to extend the target DNA strand and form QPA-PBS (primer-binding site). Then, the nicking enzyme Nt.BSTNBI was introduced to nick the target strand and release the QPA-primer-binding strand. The QPA-primer-binding strand will bind to the QPA primer and the polymerase will function to extend the strand further. The primer will then dissociate to emit fluorescence signals. The following priming step of QPA will be initiated again when the next primer binds with the QPA-primer binding strand. This method was reported to be able to reduce the background activity to allow sufficient sensitivity of the assay (Figure 5).

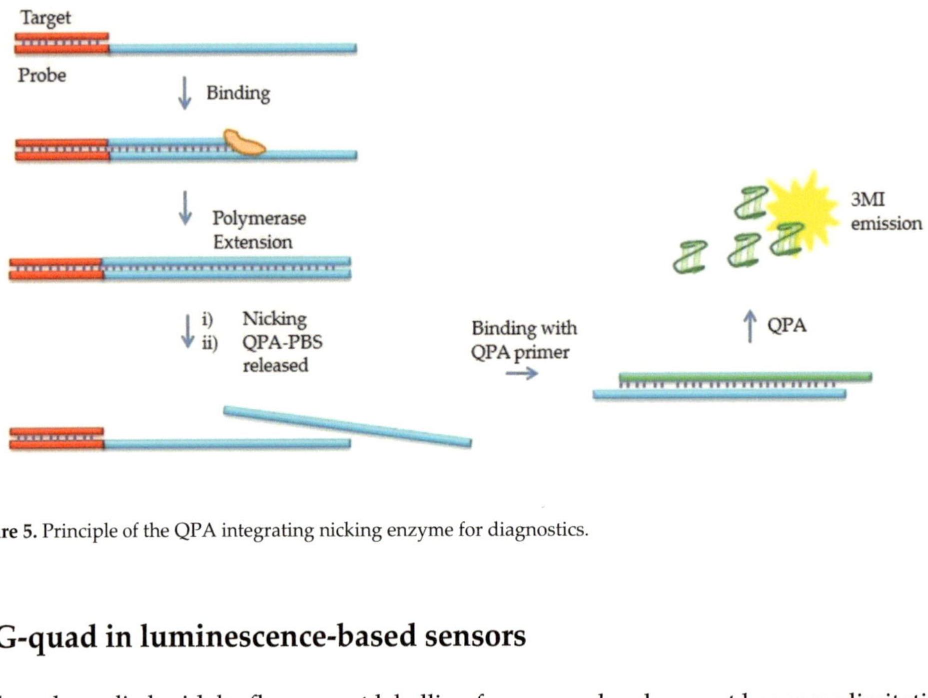

Figure 5. Principle of the QPA integrating nicking enzyme for diagnostics.

5. G-quad in luminescence-based sensors

Although applied widely, fluorescent labelling for sensor development has some limitations. Labelling the oligonucleotide covalently can reduce the binding affinity or selectivity of the oligonucleotide, which in turn hampers the efficiency of the assay. Besides, fluorescent labelling is costly and time consuming. Therefore, the use of luminescent probes in sensor development was an attractive solution. Unlike fluorescent probes, luminescent probes are not covalently attached to the nucleic acid; instead, it can bind with DNA through end stacking or electrostatic interactions, intercalation and groove binding. Such interactions will not affect the functionality of the DNA [43].

A label-free G-quad assay for metal detection was developed by Ma et al. This assay was based on the transition of a DNA structure, which involved specially designed oligonucleotide that is rich in guanine and cytosine. The transition of DNA structure would be from a quadruplex to duplex conformation induced by silver (Ag^+) ions. In this assay, platinum (II) metallo intercalator was used for the switch on/ switch off response towards the Ag^+ ions. Initially, a low luminescence background was generated due to the weak interaction of platinum (II) metallo intercalator and the G-quad structure without the Ag^+ ions. However, after adding the Ag^+ ions, cystosine-Ag^+-cytosine mismatched base pairing occurred, causing the G-quad structure to revert to a duplex structure. This allowed the intercalation of the platinum(II) metallo intercalator to the duplex DNA structure and generates a luminescent signal [44].

Besides metal detection assays, the same group also developed an assay for the detection of gene deletion (Figure 6). They designed a split G-quad assay whereby it consists of two short oligonucleotides, namely P1 and P2. Both of these strands have complementary regions that can recognize the deletion site of the chemokine receptor gene CCR5 and also contain guanine-rich overhangs. These guanine-rich overhangs of both strands can form a split G-quad in close proximity. CCR5 was used as the target because it was reported to be a co-receptor for macrophage-tropic human immunodeficiency virus type 1 (HIV-1) strain and it allows HIV-1 to enter the $CD4^+$ T cells. Mutation on CCR5 gene caused no expression of functional protein, which gives rise to the resistance against HIV infection. In this assay, P1 and P2 strands were able to hybridize with the wild-type CCR5 DNA strand. Due to the large spatial separation between both G-rich overhangs of P1 and P2, a split G-quad structure was not formed, causing the generation of a lower luminescent signal of the iridium(III) complex. However, mutant of the CCR5 DNA sequence that is shorter in length can cause the two G-quad-forming sequences of P1 and P2 to close proximity upon hybridizing with the shorter mutant sequence. This resulted in the formation of the split G-quad, generating a switch-on luminescence effect of the iridium (III) complex [45].

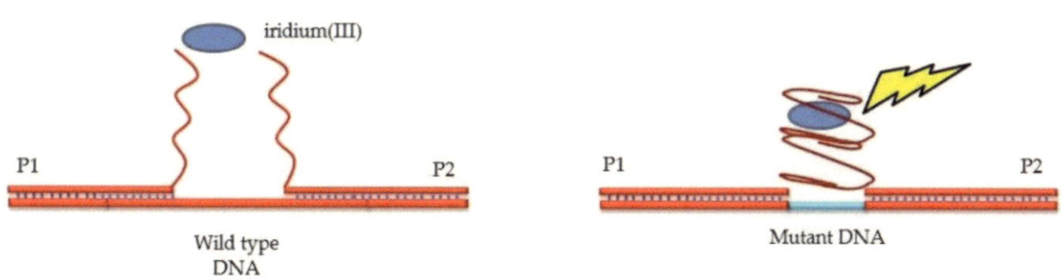

Figure 6. Split G-quad luminescent turn-on detection for the detection of gene deletion of HIV DNA sequence.

6. G-quad integrates with electrochemistry

Apart from the applications that exploit the properties of the G-quad structures, new strategies were established by integrating G-quad with electrochemistry. Recent studies developed different electrochemical G-quad sensors based on magnetic particles, nanoparticles labelled with metal tags, nanotubes and other nanomaterials. Wang et al. developed a thrombin detection assay using amplified electrochemical signal. This assay consists of an aptamer modified gold electrode and another aptamer modified with Cds hollow nanosphere. These two aptamers are complementary in most part except for the middle bases that are not complementary. This design will allow the aptamer to form a hairpin. Thus, in the presence of thrombin, the conformation of the aptamer will change from a hairpin to a G-quad structure. The nanospheres will allow the electron transfer between the gold electrode and $K_3[Fe(CN)_6]$, producing an electrocatalytic response [46].

Besides the protein detection sensor, electrochemical methods have also been exploited in the development of biosensor for cancer monitoring or impedimetric biosensor that measures the swelling behaviour of different cancer cells. An electrochemical-based G-quad sensor to detect cancer cells was developed by Qu et al. The biosensor consists of a graphene-modified electrode where aptamer, AS1411 and its complementary strand were used. In the presence of the cations, the aptamer, AS1411 forms G-quad and binds specifically to nucleolins that are expressed on the cancer cell surface with high binding affinity. Thus, only the cancer cells can be captured and it is capable to differentiate cancer cells from normal cells. The cyclic voltammetry (CV) signal of the $K_3[Fe(CN)_6]/K_4[Fe(CN)_6]$ decreased over time as the anchored aptamer folded to form a G-quad. However, after the binding of the cancer cell to the G-quad, there was little to no observable CV signal of the $K_3[Fe(CN)_6]/K_4[Fe(CN)_6]$. The sensor is then regenerated using the AS1411 complementary strand to allow it to be reused for the next round of cancer cell detection [47].

Many studies also incorporated electrochemical methods in the DNA detection sensors. Recently, Yao and co-workers integrated the isothermal exponential amplification (EXPAR) with hybridization chain reaction (HCR) of DNAzyme in addition to the merits of electrochemical method for the development of an ultrasensitive DNA sensor for avian flu strain H7N9 (Figure 7). A single-stranded DNA derived from the hemagglutinin (HA)-encoding sequences from avian influenza A (H7N9) was used as the target gene in their sensor development. The molecular beacons (MBs) contained the G-quad that cannot be opened up or assembled together without the target. A duplex probe was anchored to the surface of the electrode. Once the target DNA hybridized with one of the sequence of the probe, the amplification through toe-hole-mediated strand displacement (TMSDR) was initiated. The duplex with the target gene was released to the solution and initiated the EXPAR with a primer. On the other hand, the other bound strand of the probe on the electrode was then hybridized with the MBs. The hairpin structure of the MBs was opened through TMSDR and resulted in the formation of G-quad nanowires. Hemin was bound to the G-quad nanowires and formed DNAzyme that catalysed the oxidation of TMB, generating an increase in electrochemical signal in reduction current to be measured. This DNA sensor for avian flu (H7N9) was ultrasensitive with the limit of detection at femtomolar levels [48].

Applying electrochemical methods, the detection limits of biosensors can be improved to allow remarkably low levels to be detected. In addition to the reduced cost of the biosensor with rapid response, the ability to miniaturize the assay allows it to be considered for point-of-care applications. Such remarkable advantages have drawn many researchers to the development

http://dx.doi.org/10.5772/61580

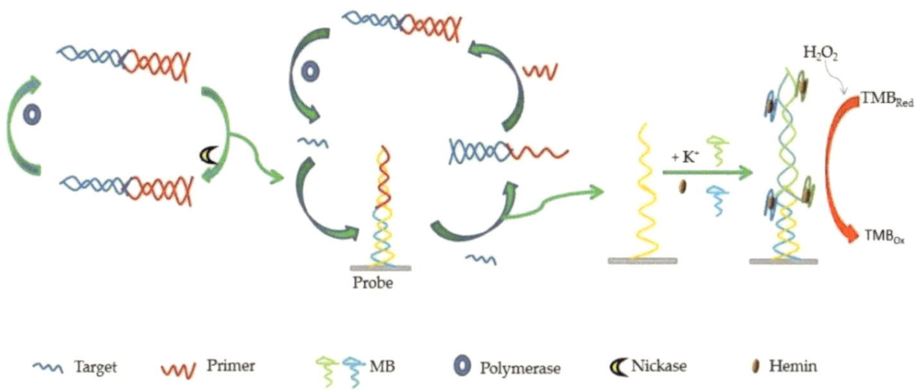

Figure 7. Schematic diagram of the DNA detection of H7N9 integrating EXPAR and HCR for electrochemical assay.

of biosensor systems based on electrochemistry for the detection of protein, DNA and other analytes.

7. G-quad integrates with quantum dots

The emergence of quantum dots (QD) in molecular sensing methods has attracted a lot of attentions from many researchers. QD have been extensively used as optical labels, probes for FRET and also as energy acceptors from metal complexes or from energy generated by luminescence or photoelectrochemical for the detection of DNA or formation of aptamer–substrate complexes [49]. Wilner and co-workers utilized the G-quad forming DNAzyme, which is conjugated to semiconductor QDs. It functions as a light source that promotes chemiluminescence resonance energy transfer (CRET) to QDs.

In this assay, anti-thrombin aptamer was used where it was also previously used in a colorimetric sensor for thrombin whereby hemin/G-quad forming DNAzyme was used to catalyse the oxidation of ABTS in the presence of H_2O_2. Meanwhile, in this assay, the hemin/G-quad forming DNAzyme aptamer–thrombin complex was used to generate chemiluminescence in the presence of H_2O_2/luminol. Then, in close proximity, this complex was able to excite the CdSe/ZnS QDs, resulting in CRET to the QDs triggering the luminescence of QDs. The glutathione-modified CdSe/ZnS QDs were attached with the anti-thrombin aptamer. A low CRET and chemiluminescence signal was observed even if it was without thrombin due to the diffusional hemin. Upon adding thrombin into the assay, the chemiluminescence signal and the CRET-stimulated luminescence of the QDs were greatly increased. The increase in CRET signals was proportional to the concentration of thrombin detected [50].

Besides the detection of aptamer–substrate complex, the detection of metal ions can also be carried out by exploiting the hemin/G-quad forming DNAzyme complex. The same group also developed a G-quad-based sensor to detect mercury ions (Hg^{2+}) (Figure 8). This assay comprised two strands of DNA whereby each of the strand has G-rich sequence forming DNAzyme (1) and (2) and a T-containing site that functions as the recognition site of Hg^{2+}, (3) and (4). As the T-containing sites of (3) and (4) were partially complementary, (1) and (2) could not assemble to form a G-quadruplex. However, in the presence of Hg^{2+}, the T-Hg^{2+}-T complexes formed between the T-containing sites of (3) and (4) and formed a duplex structure. Therefore, this led to the formation of the hemin/G-quad complex between (1) and (2). The hemin/G-quad complex catalyzed the oxidation of luminol in the presence of H_2O_2, generating chemiluminescence signal. Then, this signal acted as an internal light source that can excite CdSe/ZnS QDs, resulting in CRET to the QDs. Then, this stimulated the luminescence signal from QDs [51].

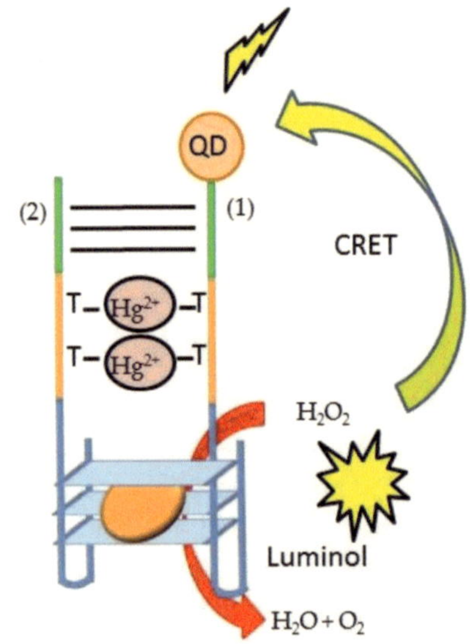

Figure 8. Detection of Hg^{2+} through CRET from luminol that is oxidized by DNAzyme to QD.

8. Conclusion

In comparison with conventional reporter systems, G-quad-based systems provide many advantages in terms of cost, thermostability and ease of synthesis. With all these advantages, it makes them very useful for the development of sensing probes for diagnosis, applicable to DNA, protein and metal detection. The vast application of G-quad structures to generate various forms of readouts ranging from colorimetric to electrochemical-based readouts puts it at the forefront of sensing reporters. G-quad-based assays are flexible as they can be adapted to many different types of diagnostic platforms. This makes G-quad an attractive alternative for the development of sensitive reporter systems for the sensing of various samples ranging from small drug molecules, chemical compounds to large biomolecules.

http://dx.doi.org/10.5772/61580

Acknowledgements

The authors would like to acknowledge funding from the Malaysian Ministry of Education under the Fundamental Research Grant Scheme (Grant No. 203/CIPPM/6711473).

Author details

Qiuting Loh and Theam Soon Lim*

*Address all correspondence to: theamsoon@usm.my

Institute for Research in Molecular Medicine, Universiti Sains Malaysia, Minden, Penang, Malaysia

References

[1] Davis JT. G-Quartets 40 years later: from 5′-GMP to molecular biology and supramolecular chemistry. Angew Chem Int Ed Engl 2004;43(6):668–98.

[2] Bochman ML, Paeschke K, Zakian VA. DNA secondary structures: stability and function of G-quadruplex structures. Nat Rev Genet. 2012;13(11):770–80.

[3] Doluca O, Withers JM, Filichev VV. Molecular engineering of guanine-rich sequences: Z-DNA, DNA triplexes, and G-quadruplexes. Chem Rev 2013;113(5):3044–83.

[4] Liu J, Cao Z, Lu Y. Functional nucleic acid sensors. Chem Rev 2009;109(5):1948–98.

[5] Ma DL, Wang M, Lin S, Han QB, Leung CH. Recent development of G-quadruplex probes for cellular imaging. Curr Top Med Chem 2015;15(19):1957–63.

[6] Lv L, Guo Z, Wang J, Wang E. G-quadruplex as signal transducer for biorecognition events. Curr Pharmaceut Design 2012;18(14):2076–95.

[7] Vummidi BR, Alzeer J, Luedtke NW. Fluorescent probes for G-quadruplex structures. Chembiochem Eur J Chem Biol 2013;14(5):540–58.

[8] Keniry MA. Quadruplex structures in nucleic acids. Biopolymers 2000;56(3):123–46.

[9] Haider SM, Neidle S, Parkinson GN. A structural analysis of G-quadruplex/ligand interactions. Biochimie 2011;93(8):1239–51.

[10] Huppert JL. Four-stranded nucleic acids: structure, function and targeting of G-quadruplexes. Chem Soc Rev 2008;37(7):1375–84.

[11] Adrian M, Heddi B, Phan AT. NMR spectroscopy of G-quadruplexes. Methods 2012;57(1):11–24.

[12] Phan AT, Kuryavyi V, Patel DJ. DNA architecture: from G to Z. Curr Opin Struct Biol 2006;16(3):288–98.

[13] Burge S, Parkinson GN, Hazel P, Todd AK, Neidle S. Quadruplex DNA: sequence, topology and structure. Nucleic Acids Res 2006;34(19):5402–15.

[14] Karsisiotis AI, O'Kane C, Webba da Silva M. DNA quadruplex folding formalism--a tutorial on quadruplex topologies. Methods 2013;64(1):28–35.

[15] Rosu F, Gabelica V, Poncelet H, De Pauw E. Tetramolecular G-quadruplex formation pathways studied by electrospray mass spectrometry. Nucleic Acids Res 2010;38(15): 5217–25.

[16] Phan AT, Kuryavyi V, Luu KN, Patel DJ. Structure of two intramolecular G-quadruplexes formed by natural human telomere sequences in K+ solution. Nucleic Acids Res 2007;35(19):6517–25.

[17] Lim KW, Amrane S, Bouaziz S, Xu W, Mu Y, Patel DJ, et al. Structure of the human telomere in K+ solution: a stable basket-type G-quadruplex with only two G-tetrad layers. J Am Chem Soc 2009;131(12):4301–9.

[18] Biffi G, Tannahill D, McCafferty J, Balasubramanian S. Quantitative visualization of DNA G-quadruplex structures in human cells. Nat Chem 2013;5(3):182–6.

[19] Schlosser K, Li Y. Biologically inspired synthetic enzymes made from DNA. Chem Biol 2009;16(3):311–22.

[20] Cuenoud B, Szostak JW. A DNA metalloenzyme with DNA ligase activity. Nature 1995;375(6532):611–4.

[21] Li Y, Breaker RR. *In vitro* selection of kinase and ligase deoxyribozymes. Methods 2001;23(2):179–90.

[22] Breaker RR, Joyce GF. A DNA enzyme that cleaves RNA. Chem Biol 1994;1(4):223–9.

[23] Li Y, Sen D. A catalytic DNA for porphyrin metallation. Nat Struct Biol 1996;3(9): 743–7.

[24] Stefan L, Xu HJ, Gros CP, Denat F, Monchaud D. Harnessing nature's insights: synthetic small molecules with peroxidase-mimicking DNAzyme properties. Chemistry 2011;17(39):10857–62.

[25] Saito K, Tai H, Hemmi H, Kobayashi N, Yamamoto Y. Interaction between the heme and a G-quartet in a heme-DNA complex. Inorg Chem 2012;51(15):8168–76.

[26] Kosman J, Juskowiak B. Peroxidase-mimicking DNAzymes for biosensing applications: a review. Anal Chim Acta 2011;707(1–2):7–17.

[27] Stefan L, Denat F, Monchaud D. Insights into how nucleotide supplements enhance the peroxidase-mimicking DNAzyme activity of the G-quadruplex/hemin system. Nucleic Acids Res 2012;40(17):8759–72.

[28] Pelossof G, Tel-Vered R, Elbaz J, Willner I. Amplified biosensing using the horseradish peroxidase-mimicking DNAzyme as an electrocatalyst. Anal Chem 2010;82(11): 4396–402.

[29] Golub E, Freeman R, Niazov A, Willner I. Hemin/G-quadruplexes as DNAzymes for the fluorescent detection of DNA, aptamer-thrombin complexes, and probing the activity of glucose oxidase. The Analyst 2011;136(21):4397–401.

[30] Freeman R, Sharon E, Teller C, Henning A, Tzfati Y, Willner I. DNAzyme-like activity of hemin-telomeric G-quadruplexes for the optical analysis of telomerase and its inhibitors. Chembiochem Eur J Chem Biol 2010;11(17):2362–7.

[31] Zhou X-H, Kong D-M, Shen H-X. G-quadruplex–hemin DNAzyme-amplified colorimetric detection of Ag^+ ion. Anal Chim Acta 2010;678(1):124–7.

[32] Omar N, Loh Q, Tye G, Choong Y, Noordin R, Glökler J, et al. Development of an antigen-DNAzyme based probe for a direct antibody-antigen assay using the intrinsic DNAzyme activity of a Daunomycin Aptamer. Sensors 2013;14(1):346–55.

[33] Li R, Xiong C, Xiao Z, Ling L. Colorimetric detection of cholesterol with G-quadruplex-based DNAzymes and ABTS2. Anal Chim Acta 2012;724:80–5.

[34] Yang L, Du F, Chen G, Yasmeen A, Tang Z. A novel colorimetric PCR-based biosensor for detection and quantification of hepatitis B virus. Anal Chim Acta 2014;840:75–81.

[35] Lin X, Chen Q, Liu W, Li H, Lin J-M. A portable microchip for ultrasensitive and high-throughput assay of thrombin by rolling circle amplification and hemin/G-quadruplex system. Biosensors Bioelectron 2014;56:71–6.

[36] Loh Q, Omar N, Glökler J, Lim TS. IQPA: Isothermal nucleic acid amplification-based immunoassay using DNAzyme as the reporter system. Anal Biochem 2014;463:67–9.

[37] Miannay F-A, Gustavsson T, Banyasz A, Markovitsi D. Excited-state dynamics of dGMP measured by steady-state and femtosecond fluorescence spectroscopy†. J Phys Chem A. 2010;114(9):3256–63.

[38] Fu L, Li B, Zhang Y. Label-free fluorescence method for screening G-quadruplex ligands. Anal Biochem 2012;421(1):198–202.

[39] Monchaud D, Allain C, Teulade-Fichou MP. Thiazole orange: a useful probe for fluorescence sensing of G-quadruplex-ligand interactions. Nucleosides Nucleotides Nucleic Acids 2007;26(10–12):1585–8.

[40] Yang H, Cui H, Wang L, Yan L, Qian Y, Zheng XE, et al. A label-free G-quadruplex DNA-based fluorescence method for highly sensitive, direct detection of cisplatin. Sensors Actuators B: Chemical 2014;202:714–20.

[41] Li T, Dong S, Wang E. A Lead(II)-Driven DNA Molecular Device for Turn-On Fluorescence Detection of Lead(II) Ion with High Selectivity and Sensitivity. Journal of the American Chemical Society. 2010;132(38):13156-7.

[42] Gogichaishvili S, Lomidze L, Kankia B. Quadruplex priming amplification combined with nicking enzyme for diagnostics. Anal Biochem 2014;466:44–8.

[43] He HZ, Chan DS, Leung CH, Ma DL. G-quadruplexes for luminescent sensing and logic gates. Nucleic Acids Res 2013;41(8):4345–59.

[44] Man BY, Chan DS, Yang H, Ang SW, Yang F, Yan SC, et al. A selective G-quadruplex-based luminescent switch-on probe for the detection of nanomolar silver(I) ions in aqueous solution. Chem Comm (Cambridge, England). 2010;46(45):8534–6.

[45] He HZ, Chan DS, Leung CH, Ma DL. A highly selective G-quadruplex-based luminescent switch-on probe for the detection of gene deletion. Chemical Commun (Cambridge, England) 2012;48(76):9462–4.

[46] Li Y, Bao J, Han M, Dai Z, Wang H. A simple assay to amplify the electrochemical signal by the aptamer based biosensor modified with CdS hollow nanospheres. Biosensors Bioelectron 2011;26(8):3531–5.

[47] Feng L, Chen Y, Ren J, Qu X. A graphene functionalized electrochemical aptasensor for selective label-free detection of cancer cells. Biomaterials. 2011;32(11):2930–7.

[48] Yu Y, Chen Z, Jian W, Sun D, Zhang B, Li X, et al. Ultrasensitive electrochemical detection of avian influenza A (H7N9) virus DNA based on isothermal exponential amplification coupled with hybridization chain reaction of DNAzyme nanowires. Biosensors Bioelectron 2015;64:566–71.

[49] Zhang Y, Wang TH. Quantum dot enabled molecular sensing and diagnostics. Theranostics 2012;2(7):631–54.

[50] Liu X, Freeman R, Golub E, Willner I. Chemiluminescence and chemiluminescence resonance energy transfer (CRET) aptamer sensors using catalytic hemin/G-quadruplexes. ACS Nano 2011;5(9):7648–55.

[51] Freeman R, Liu X, Willner I. Chemiluminescent and chemiluminescence resonance energy transfer (CRET) detection of DNA, metal ions, and aptamer–substrate complexes using hemin/G-quadruplexes and CdSe/ZnS quantum dots. J Am Chem Soc 2011;133(30):11597–604.

Chapter 8

Small-molecule Nucleic-acid-based Gene-silencing Strategies

Zhijie Xu and Lifang Yang

Additional information is available at the end of the chapter

http://dx.doi.org/10.5772/62137

Abstract

Gene-targeting strategies based on nucleic acid have opened a new era with the development of potent and effective gene intervention strategies, such as DNAzymes, ribozymes, small interfering RNAs (siRNAs), antisense oligonucleotides (ASOs), aptamers, decoys, etc. These technologies have been examined in the setting of clinical trials, and several have recently made the successful transition from basic research to clinical trials. This chapter discusses progress made in these technologies, mainly focusing on Dzs and siRNAs, because these are poised to play an integral role in antigene therapies in the future.

Keywords: Gene-targeting strategies, DNAzymes, siRNAs, basic research, clinical trials

1. Introduction

Over the past decade, it is known that the advent of oligonucleotide-based gene inactivation agents have provided potential for these to serve as analytical tools and potential treatments in a range of diseases, including cancer, infections, inflammation, etc. During this time, many genes have been targeted by specifically engineered agents from different classes of small-molecule nucleic-acid-based drugs in experimental models of disease to probe, dissect, and characterize further the complex processes that underpin molecular signaling. Subsequently, a number of molecules have been examined in the setting of clinical trials, and several have recently made the successful transition from the bench to the clinic, heralding an exciting era of gene-specific treatments. This is particularly important because clear inadequacies in present therapies account for significant morbidity, mortality, and cost. The broad umbrella of gene-silencing therapeutics encompasses a range of agents that include deoxyribozymes (DNAzymes, Dzs), ribozymes, siRNAs, ASOs, aptamers, and decoys. This chapter tracks current movements in these technologies, focusing mainly on Dzs and siRNAs, because these are poised to play an integral role in antigene therapies in the future.

2. DNAzymes

Among the gene-silencing technologies, Breaker and Joyce, in 1994, used an *in vitro* selection method to identify a special Dz from a random pool of single-stranded DNA to catalyze Pb^{2+}-dependent cleavage of an RNA phosphodiester linkage [1]. Afterward, a number of Dzs were created with the capacity to catalyze many reactions, including the cleavage of DNA or RNA, the modification and ligation of DNA, and the metalation of porphyrin rings. However, because of the low efficiency of RNA cleavage, they are not widely used for biological applications except for 10-23 Dz [2]. The inherent catalytic RNA-cleaving property of Dzs has been used with different mRNA targets as *in vitro* diagnostic and analytical tools, as well as *in vivo* therapeutic agents.

2.1. The possible mechanisms and characteristics of DNAzymes

Dzs of the 10-23 subtype are single-stranded DNA catalysts that comprise a central cation-dependent catalytic core of around 15 deoxyribonucleotides [ggctagctacaacga], and two complementary binding arms of 6–12 nucleotides that are specific for each site along the target RNA transcript [3]. As diagrammed in Figure 1, the enzyme binds the substrate through Watson-Crick base pairing and cleaves a particular phosphodiester linkage located between an unpaired purine and paired pyrimidine in the RNA. This results in the formation of 5′ and 3′ products, which contain a 2′, 3′-cyclic phosphate and 5′-hydroxyl terminus, respectively. Even though the 10-23 Dz can cleave any RY junction, the reactivity of each substrate dinucleotide compared in the same background sequence with the appropriately matched DNAzyme is found to follow the scheme AU = GU > GC >> AC. Murray *et al.* found that when the target site core is an RC dinucleotide, the relatively poor activity could be enhanced up to 200-fold by substituting deoxyguanine with deoxyinosine, which could effectively reduce the strength of Watson-Crick pairing between bases flanking the cleavage site [4].

Due to the simple cleavage-site requirement, Dzs are capable of cleaving any particular mRNAs for multiple turnover by appropriately designing the sequence in the binding arms. Several features make Dzs attractive from a drug developmental viewpoint. For example, these are inexpensive to synthesize, and their small size allows specificity. Moreover, DNAzymes can be rendered more stable by structural modifications, such as phosphorothioate (PS) linkages, locked nucleic acids (LNAs), and 3′-3′ inverted nucleotide end of the DNAzyme [5]. Enhanced biostability, low toxicity, affinity, and versatility suggest great promise for diagnostic and therapeutic applications [6]. Limitations thus far in the development of DNAzymes as novel therapeutics have been delivery and biodistribution, which revolve around poor cellular uptake and stability. Delivery systems depend on the route of administration and the target site. Moreover, an ideal delivery system would facilitate rapid and efficient distribution to the site of action, stability, low toxicity, and efficacy.

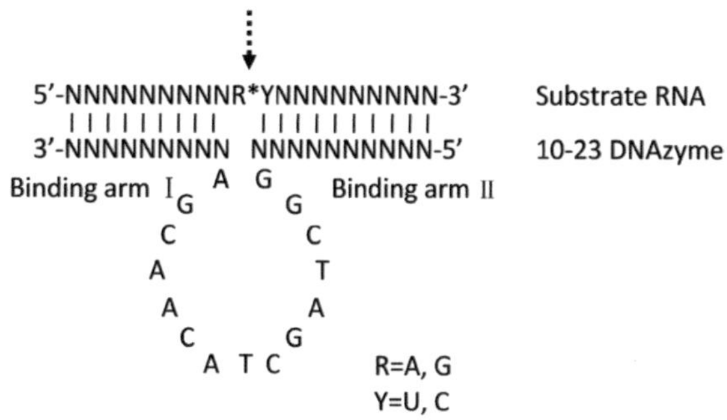

Figure 1. Secondary structure of the 10-23 DNAzyme–substrate complexes. The 10-23 DNAzyme consists of two variable binding arms, designated arm I and arm II, which flank a conserved 15 base unpaired motif that forms the catalytic core. The only requirement of the RNA substrate is for a core sequence containing an RY junction.

2.2. DNAzymes delivery systems – Past to present

As in all nucleic-acid-based reagents, efficient drug delivery systems (DDSs) to deliver the Dzs to targeting site are highly needed. Furthermore, by adopting DDSs, it could be helpful to solve the obstacles about DNAzymes' stability, biological effects, and toxicity. Several seminal studies have demonstrated that certain DNAzyme delivery systems can efficiently encapsulate DNAzymes and transfect them into cells without clear toxicity. The attempt first involved the microspheres of co-polymers poly (lactic acid) and poly (glycolic acid) (PLGA), which encapsulated the Dzs. PLGA microspheres are able to achieve biphasic release and sustained accumulation of the Dzs [7]. In a second delivery system, a chimeric aptamer–DNAzyme conjugate was generated for the first time using a nucleolin aptamer (NCL-APT) and survivin Dz (Sur_Dz). This conjugate could be used as a specific gene-targeting therapy to kill the targeted cancer cells [8]. A third delivery system is developed and studied based on the cationic liposomal formulation technology. Li *et al.* reported the effect of a c-Jun targeted DNAzyme (Dz13) in a rabbit model of vein graft stenosis after autologous transplantation in a cationic liposomal formulation containing 1,2-dioleoyl-3-trimethylammonium propane (DOTAP)/1,2-dioleoyl- snglycero-3-phosphoethanolamine (DOPE). Dz13/DOTAP/DOPE allows sufficient uptake by the veins and reduces SMC (smooth muscle cell) proliferation and c-Jun protein expression *in vitro*. Meanwhile, a Phase I clinical trial has indicated that it is safe and well tolerated after local administration in skin cancer patients [9]. Finally, due to their low toxicity and no side effects, nanoparticulate systems have spread rapidly and could significantly enhance the efficacy of tumoricidal Dzs. Marquardt *et al.* found that c-Jun targeted Dz13 delivered in this manner is capable of enhanced skin penetration efficiency and cellular uptake with a high reduced degradation of Dz13 in *vitro* [10]. These results indicate that, with more suitable delivery approaches, the biological effects of Dzs would be further increased, and the Dzs could be applied to new subject areas.

2.3. Application of DNAzymes *in vivo* and *in vitro*

Increasing evidence indicates the efficacy and potency of DNAzymes *in vivo* and *in vitro* in a range of disease settings, allowing characterization of key pathogenic pathways and their potential use as therapeutic agents (Table 1). DNAzymes have been widely applied as a new interference strategy in the treatment of many conditions, including cancer, viral diseases, and vein graft stenosis. For instance, Dz13 targeting the transcription factor c-Jun has shown promise in experimental models of mice infected with H5N1 virus via reducing H5N1 influenza virus replication and decreasing expression of pro-inflammatory cytokines [11]. Furthermore, Dz13/DOTAP/DOPE reduces SMC proliferation and c-Jun protein expression *in vitro*, and inhibits neintima formation after end-to-side transplantation, which may potentially be useful to reduce graft failure [9]. Likewise, Cai *et al.* demonstrated that safe and well-tolerated Dz13 could inhibit tumor growth and reduce lung nodule formation in a model of metastasis [12].

Target	Summary Description on Biological Effects (In Vitro and In Vivo)	Refs.
LMP1	·Inhibiting proliferation and metastasis ·Promoting apoptosis ·Enhancing radiosensitivity	[13-15]
Egr-1	·Inhibiting proliferation and metastasis ·Suppressing tumor growth	[16]
MMP-9	·Inhibiting invasion and metastasis ·Suppressing tumor growth	[17, 18]
IGF-II	·Inhibiting proliferation ·inducing caspase-dependent apoptosis	[19]
survivin	·Inhibiting proliferation ·Promoting apoptosis	[8]
β-integrin	·Inhibiting invasion and metastasis ·Blocking angiogenesis	[20]
VEGFR-1	·Blocking angiogenesis ·Suppressing tumor growth	[21]
DNMT1	·Inhibiting proliferation	[22]
Bcl-XL	·Promoting apoptosis ·Enhancing Taxol chemosensitivity	[23]
c-Jun	·Inhibiting proliferation ·Restraining virus replication and host inflammation ·Suppressing tumor growth	[9, 11, 12]
BCR-ABL T315I	·Overcoming imatinib resistance based on BCR-ABL T315I Mutation	[24]
EGFR T790M	·Overcoming EGFR T790M mutant-based TKI resistance	[25]
TXNIP	·Attenuating oxidative stress, renal fibrosis, and collagen deposition	[26]

Table 1. *In vivo* and *in vitro* applications of 10-23 DNAzymes

As is well known, treatment resistance is one of the leading causes of tumor recurrence. We have recently evaluated Dz1 targeting latent membrane protein 1 (LMP1) in the setting of nasopharyngeal carcinoma model and demonstrated that injected intratumorally DZ1 with fuGENE 6 in nude mice inoculating LMP1-positive cells resulted in a significant inhibition of tumor growth and an enhanced radiosensitivity. Dynamic contrast-enhanced magnetic resonance imaging (DCE-MRI) showed that DZ1 reduces the angiogenesis and microvascular permeability [13]. Other studies have used DNAzymes to target the other key genes in cancer therapy. DNAzyme targeting the Bcl-XL gene significantly sensitized a panel of cancer cells to apoptosis and further to reverse the chemoresistant phenotype [23]. Due to a secondary mutation at T790M in the epidermal growth factor receptor (EGFR), most of nonsmall-cell lung cancer (NSCLC) patients will eventually develop resistance to tyrosine kinase inhibitors (TKIs) treatment. Allele-specific silencing of EGFR T790M expression and downstream signaling by DNAzyme DzT could suppress the growth of xenograft tumors derived from H1975$^{TM/LR}$ cells, indicating that DzT is capable of overcoming EGFR T790M mutant-based TKI resistance [25]. In a similar way, Kim *et al.* developed the DNAzyme that specifically targets the site of the point mutation (T315I), conferring imatinib resistance in BCR–ABL mRNA. Cleavage of T315I-mutant ABL mRNA by DNAzyme could significantly induce apoptosis and inhibit proliferation in imatinib-resistant BCR-ABL-positive cells [24].

2.4. DNAzymes in clinical trials

The favorable properties of 10-23 Dzs, such as their enhanced biological stability, negligible side effects, and lack of immunogenicity, have paved the way for Dzs to enter clinical trials [17]. Up to now, Dzs to three targets have been undergoing clinical trials and at least one of them has proved its therapeutic efficacy in Phase II trials (Table 2). These results further show the potential of Dzs therapeutic approach for the treatment of diseases and represent a major advance in this field.

As we have found that LMP1-targeted Dz1 could effectively inhibit the growth and enhance the radiosensitivity of NPC cells both *in vivo* and *in vitro*, we investigated the antitumor and radiosensitizing effects of Dz1 in NPC patients for the first time [27]. Being safe and well tolerated, a randomized and double-blind clinical study was conducted in 40 NPC patients, who received Dz1 or saline intratumorally in conjunction with radiation therapy. In a 3-month follow-up, compared with the saline control group, the mean tumor regression and undetectable EBV-DNA copy number in the DZ1 group is significantly higher. Molecular imaging analysis found that Dz1 was tested to accelerate the decline of K^{trans}, generally recognized as a marker of tumor blood flow and permeability [28].

The nuclear transcription factor c-Jun is preferentially expressed in a range of cancers. Dz13 cleaves at the G1311U junction in human c-jun mRNA and exerts its antitumor activity via induction of apoptosis, inhibition of angiogenesis, and the induction of adaptive immunity [11]. A phase I first-in-human trial is conducted to determine the safety and tolerability of Dz13 in nine patients with basal-cell carcinoma (BCC), who received a single intratumoral injected

Target	Disease	Phase	Trial ID	Refs.
LMP1	Nasopharyngeal carcinoma	Phases I/II Completed	NCT01449942	[27, 28]
c-Jun	Nodular basal-cell carcinoma Melanoma with satellite or in-transit metastasis	Phases I Completed Phase I/Ib Ongoing	ACTRN12610000162011 ACTRN12613000302752	[29]
GATA-3	Asthma	Phases I Completed	NCT01470911	[30-33]
	Atopic dermatitis	Phases I Completed Phases I Completed Phases II Completed Phase IIa Completed Phases I Completed	NCT01554319 NCT01577953 NCT01743768 EUCTR2012-003570-77-DE NCT02079688	
		Phases IIa Ongoing		
	Ulcerative colitis Chronic obstructive pulmonary disease	Phases I/II Ongoing Phase IIa Pending	NCT02129439 DRKS00006087	
	Atopic eczema	Phase IIa Ongoing	EUCTR2013-001091-38-DE	

Table 2. Clinical trials of DNAzymes in anti-diseases therapy

dose of Dz13 (10, 30, or 100 ìg) [29]. Followed-up over four weeks, c-Jun expression is reduced in all nine participants. Meanwhile, Dz13 could significantly promote apoptosis and stimulate inflammatory and adaptive immune responses in the tumors. Among the participants, five patients have a reduction in histological tumor depth. These results indicated that Dz13 possibly could represent a future treatment option for BCC prior to excision by surgery.

The transcription factor GATA-3 plays an important role in the regulation of Th2-mediated immune mechanisms such as in allergic bronchial asthma, and the DNAzyme hgd40 has been shown to specifically and selectively reduce expression of GATA-3 mRNA. Turowska *et al.* found that hgd40 is evenly distributed in inflamed asthmatic mouse lungs within minutes after single dose application, and could slowly eliminate from lung tissue with the goal to minimize accumulation and to ensure continued exposure for efficacy [32]. Safety pharmacology studies showed that with no observable adverse event, hgd40 has a highly favorable toxicity profile when administered by aerosol inhalation at the therapeutic doses [33]. With good safety and tolerability in the phase I program [31], a randomized, double-blind, placebo-controlled, multicenter clinical trial of hgd40 was conducted in patients with allergic asthma, who had biphasic early and late asthmatic responses after laboratory-based allergen provocation [30]. After each study drug administered by inhalation once daily for 28 days, hgd40 significantly attenuates both late and early asthmatic responses and improves lung function. Moreover, the Th2-regulated inflammatory responses are also attenuated.

http://dx.doi.org/10.5772/62137

These studies, taken together, further demonstrate the potential use of DNAzymes as gene-targeting drugs. As Dzs are safe and well tolerated in humans, there is a good chance that we may witness the Dzs reaching the clinic in the near future.

3. Small interfering RNA

Small interfering RNA (siRNA), first discovered in plants and *Caenorhabditis elegans* and later in mammalian cells, is a member of a family of noncoding RNAs (ncRNAs) that affect and regulate gene transcriptional and posttranscriptional silencing [34]. This sequence-specific gene-silencing phenomenon could cause mRNA to be effectively broken down after transcription, resulting in no obvious translation. SiRNA represents an emerging therapeutic approach against diseases for *in vivo* and *in vitro* studies, and along with novel drug delivery techniques, the challenge of siRNA-based therapeutics is only now being optimized. These discoveries led to a surge in interest in harnessing siRNA for biomedical research and drug development.

3.1. The possible mechanisms and challenges of siRNAs

SiRNAs, synthetic mediators of RNA interference (RNAi), are basically dsRNA molecules designed specifically to silence expression of target genes. Cytoplasmic dsRNA molecules are considered unusual and are substrate for endonuclease Dicer, an RNase III family member. Vertebrate-specific TAR (HIV trans-activator RNA) RNA-binding protein (TRBP) and protein kinase R-activating protein (PACT) help Dicer to identify and dice dsRNA into about 21 bp fragments with 2 nucleotides overhangs at each end, generating the siRNA. Then recognized by an important enzyme Argonaute 2 (AGO2), siRNA of 21-23 nucleotides are incorporated into an RNA-induced silencing complex (RISC). RNA helicases unwind the double-stranded siRNA. The sense strand of the double-stranded siRNA is cleaved during the formation of the RISC complex, and the antisense strand guides RISC to the complementary target mRNA, which is rapidly degraded by RISC (Figure 2) [35, 36].

Though siRNAs can efficiently silence target gene expression in a sequence-specific manner, many challenges, including rapid degradation, poor cellular uptake, off-target effects and immune response, need to be addressed in order to carry these molecules into clinical trials [37, 38]. For example, Chung *et al.* illustrated the underappreciated off-target effects of siRNA gene knockdown technology. Hepatitis C Virus (HCV) depends on a core MOBKL1B (Mps one binder kinase activator-like 1B)–NS5A peptide complex to complete its life cycle. However, without the absence of MOBKL1B, siRNA of MOBKL1B still has off-target inhibitory effects on virus replication [39]. Researchers have tried to develop modified method to reduce the disadvantages. By using the default parameters in siDirect 2.0 Web server (http://siDirect2.RNAi.jp/), at least one qualified siRNA for >94% of human mRNA sequences in the RefSeq database can be designed [40]. In addition, chemical modifications have been shown to protect siRNAs from nuclease degradation without interfering with siRNA-silencing efficiency [37]. Thus, improvements in rational design strategies might have the potential to make the siRNAs

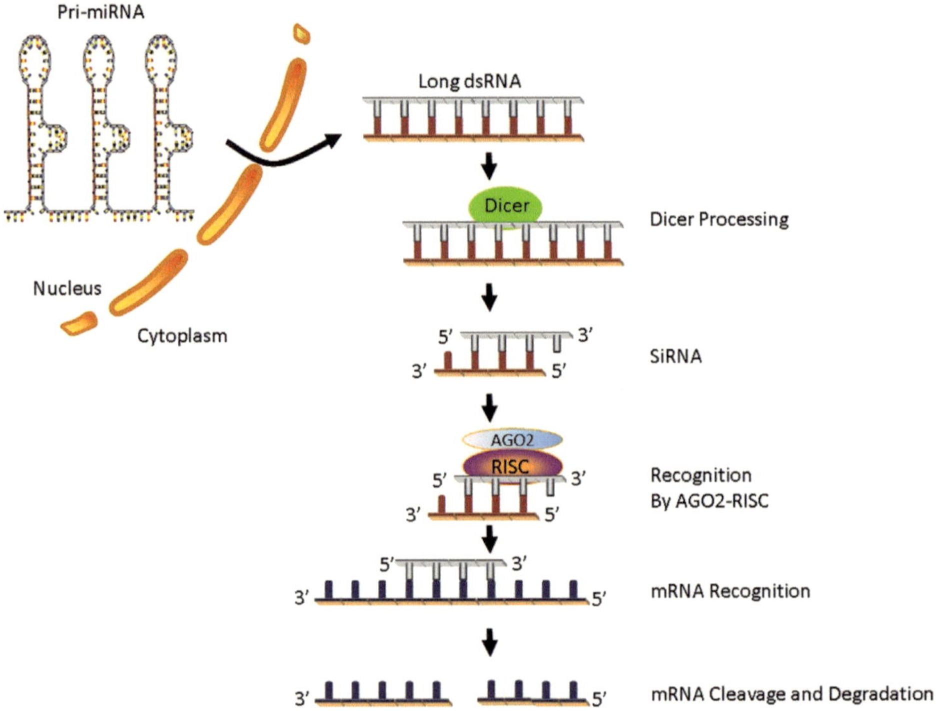

Figure 2. The process of siRNA-mediated degradation of target mRNA in eukaryotic cells. siRNA is recognized by AGO2 and incorporated into the RISC. After that, RNA helicases unwind the double-stranded siRNA, and the antisense strand guides RISC to the complementary targeted mRNA, which is cleaved by RISC and rapidly degraded.

more effective in the near future and to open the door to development of highly effective and safe therapeutics for clinical applications.

3.2. SiRNAs delivery systems

Delivery of siRNAs to target tissues is impeded by many barriers at different levels. As possible drugs in the near future, targeted delivery of siRNAs provides remarkable opportunities for accelerating RNAi-based high-performance treatments. The success of siRNAs-based delivery systems may be dependent upon uncovering a delivery route and sophisticated delivery carriers. In this regard, Fujita *et al.* have reported a powerful platform (PnkRNA™ and nkRNA®) to promote naked RNAi approaches through inhalation without delivery vehicles in lung cancer xenograft models. This modified local drug delivery system could offer a promising strategy for enhancing RNAi effects in cancer therapy [41]. In addition, with high binding specificity, nucleic acid aptamer represents a different promising tool for selective delivery siRNAs to cancer cells or tissues, resulting in increasing the therapeutic efficacy as well as reducing toxicity [42]. Likewise, the latest studies in using cell-penetrating peptides (CPPs) combined with molecular cargos, including liposomes, polymers, nanoparticles, and so on, have indicated that for the delivery of siRNAs, the combination strategy can remit the reduced internalization efficiency caused by neutralization [43]. However, each transfection process needs to be optimized because of cell density, siRNA concentration, transfection reagents, etc.

3.3. Application of siRNAs – From the bench to the clinic

The discovery of RNA interference (RNAi) was approximately 20 years ago, and opened up a new mechanism for gene-silencing therapeutics. Kim *et al.* evaluated the inhibition effect on Notch1 expression by siRNA, and found that Notch1-targeted-siRNA could result in retarded progression of inflammation, bone erosion, and cartilage damage in collagen-induced arthritis (CIA) mice by efficiently inhibiting the expression of Notch1 in mRNA level [44]. Cao *et al.* demonstrated that after silencing the expression of vascular endothelial growth factor (VEGF) by siRNA, the number of living cells on the gel and the mucosa thickness are significantly decreased *in vivo*, which indicated siRNA-targeting VEGF may be useful as a convenient therapeutic option for chronic rhinosinusitis [45]. Similarly, VEGF-siRNA decreases the vessel-forming ability and exhibited no testable cytotoxicity by significantly decreasing the expression of VEGF mRNA and protein [46].

To date, given the progress of basic research, there are examples of clinical trial projects based on RNAi technology against cancer and other diseases. SiRNA therapeutics is now well poised to enter the clinical formulary as a new class of drugs in the near future. In an open-label phase I/IIa study in the first-line setting of fifteen patients with nonoperable locally advanced pancreatic cancer (LAPC), an siRNA drug (G12D) against KRAS, a Kirsten ras oncogene homolog from the mammalian ras gene family, is well tolerated, safe, and demonstrated a potential therapeutic efficacy to the patients enrolled, when combined with chemotherapy. However, five participants experienced serious adverse events [47]. In addition, a recent systematic analysis of a new RNAi therapeutic agent based on cationic lipoplexes containing chemically stabilized siRNAs, called Atu027, which silences expression of protein kinase N3 in the vascular endothelium in patients with advanced solid tumors. In one case of 24 patients, the study showed that Atu027 is tolerated up to 0.180 mg/kg, and no obvious dose-dependent toxicities are observed [48]. Likewise, the results from another case of 34 patients showed that Atu027 is safe in patients with advanced solid tumors, with 41% of patients having stable disease for at least 8 weeks [49]. Also, because SYL040012 is an siRNA designed to specifically silence β adrenergic receptor 2 (ADRB2) currently under development for glaucoma treatment *in vivo* and *in vitro* [50], a phase I clinical trial of SYL040012 with 30 healthy subjects having intraocular pressure (IOP) below 21 mmHg was conducted [51]. This trial found that administration of SYL040012 over a period of 7 days significantly reduced IOP values regardless of the dose used, was well tolerated locally and had no local or systemic adverse events. Thus, taken together, these clinical studies conducted on siRNAs in the past few years indicate that safe and effective target gene knockdown is achievable. Though targeting any individual gene might lead to unanticipated clinical toxicity that could stop the development of any individual siRNA drug, we anticipate a rapid expansion of clinical trials for multiple clinical indications.

4. Antisense oligonucleotides

Antisense oligonucleotide, first recognized in 1978 by Zamecnik and Stevenson, is a small synthetic piece of DNA (usually 15–18 mer in length) that can bind complementary RNA by Watson-Crick base pairing. ASOs can target most RNA transcripts and have emerged as the ideal therapeutic agents for a broad number of diseases [52, 53]. Upon binding to their target, ASOs can modulate the intermediary metabolism of RNA by the recruitment of endogenous RNase H1 to interfere with RNA function [54]. Human RNase H1 is a ubiquitous enzyme that hydrolyzes the duplex formed between a DNA containing ssASO and target RNA through its N-terminus RNA-binding domain. In order to cleave the RNA in the duplex, the RNase H1 catalytic domain needs at least 5 consecutive DNA/RNA base pairs, and cleavage usually occurs within 7–10 nucleotides from the 5′-end of the RNA. After cleavage, the exposed phosphate on the 5′-end and hydroxyl on the 3′-end are recognized, and the RNA is subsequently degraded by cellular nucleases. At some point after RNase H1 cleaves the RNA, the ssASO is released and is available to reengage another transcript.

Even though much progress has been made in the ASO field so far, there are still many questions that might result in nonspecific effects. One of the principle challenges for success is efficacious delivery to target organs. Because initial ASO molecules are either of low affinity or low membrane permeability, they suffered from poor solubility and rapid degradation by nucleases. In the field, many studies to improve the therapeutic potential of ASOs have focused on chemical modifications to either improve nuclease resistance, such as 2′-O-methoxyethyl (2′-MOE), or to facilitate cellular uptake, like phosphorothioate backbone that improves membrane penetration [55, 56]. Moreover, too many heparin-binding cell surface proteins have been identified to bind the phosphorothioate oligo with nanomolar affinity. The delivery of ASO drug, encapsulating with materials ranging from cationic lipids to dendrimers to alginate/chitosan nanoparticles, has reached new heights of clinical acceptance [52].

Over the past several years, antisense oligonucleotide-based targeted therapy has emerged rapidly. Interest in the field has ramped-up dramatically, as numerous ongoing clinical trials are evaluating the treatment effect on diseases with ASOs. Antisense oligonucleotide sodium LY2181308 (LY2181308), hybridizing to the human survivin mRNA, is well tolerated in patients with acute myeloid leukemia (AML). In combination with chemotherapy, LY2181308 does not cause additional toxicity, though 1/16 patients had incomplete responses, and 4/16 patients had cytoreduction [57]. Thus, future clinical trials are needed to further confirm its clinical benefit. In another open-label, parallel-group study, reducing factor XI levels by a second-generation antisense oligonucleotide FXI-ASO (ISIS 416858) is an effective method for prevention of postoperative venous thromboembolism. With respect to the risk of bleeding, FXI-ASO received once daily appeared to be safe [58]. In another phase II trial, compared with those who received placebo, the participants with Crohn's disease who received SMAD7 ASO Mongersen (formerly GED0301) had significantly higher rates of remission and clinical response [59]. Even more important, mipomersen, an antisense agent targeted to apolipoprotein B, has recently received FDA (United States Food and Drug Administration) approval for the treatment of familial hypercholesterolemia (http://www.fda.gov/newsevents/newsroom/

pressannouncements/ucm337195.htm). This compelling therapeutic potential powerfully supports further clinical investigations of ASOs in subjects in the near future.

5. Ribozymes

Ribozymes, also termed catalytic RNA, are highly structured RNA sequences that can be engineered to specifically cleave target RNA molecules, similar to the action of protein enzymes. However, unlike protein ribonucleases, ribozymes cleave only at a specific location, using base-pairing and tertiary interactions to help align the cleavage site within the catalytic core. The general mechanism of ribozymes is as follows: a 2′-oxygen nucleophile attacks the adjacent phosphate in the target RNA backbone, resulting in cleavage products with 2′, 3′-cyclic phosphate and 5′ hydroxyl termini [60].

Since ribozymes were accidentally discovered in 1982, it has been shown that RNA can act in at least two ways in biology: as genetic material and as a biological catalyst. Examples of ribozymes include the hammerhead ribozyme, the Leadzyme, and the hairpin ribozyme. In the last several years, crystal structures of these ribozymes have been determined, providing detailed views of the tertiary folds of these RNAs [60, 61], which would be modulated allosterically to increase specificity of ribozyme action.

Compared to other therapeutical RNAs such as siRNAs, the current therapeutic efficacy of ribozymes remains low due to their limited specificity, and structural instability [62]. And furthermore, the amount of free Mg^{2+} in the intracellular environment plays a critical limitation role for the catalytic activity [63]. To date, gene-therapy-based studies have focused upon developing strategies to stabilize ribozymes and transfect them into live cells. Rouge *et al.* reported the concept of ribozyme-spherical nucleic acid (SNA) conjugates and found that these conjugates could allow high cellular uptake of ribozymes, with favorable catalytic activity and stability [64]. Paudel et al. studied the effect of molecular crowding agents, like polyethylene glycol (PEG), on the folding and catalysis of ribozymes. They demonstrated that PEG favors the formation of the docked structure, which increases ribozymes' activity. In addition, Mg^{2+}-induced folding in the presence of PEG occurs at concentrations ~ 7-fold lower than in the absence of PEG [65].

Up to now, at least two clinical trials have positively showed the safety, feasibility, and long-term stability of using ribozymes targeted to different mRNAs, such as HIV (human immunodeficiency virus) elements [66] and VEGF-1 [67]. However, the transduction efficiency left room for improvement. In a phase II cell-delivered gene transfer clinical trial, 74 HIV-1 infected adults enrolled randomly received a tat/vpr specific ribozyme OZ1 or placebo. This study showed that OZ1-based gene therapy is safe, and has modest efficacy. In the future, modifications would aim to increase the lymphocyte recovery in order to enhance the therapeutic effect [68]. Another phase II trial of RPI.4610, an antiangiogenic ribozyme targeting the VEGFR-1 mRNA, also demonstrated a well-tolerated safety profile but lacked the clinical efficacy, which results in precluding this drug from further development [69]. Thus, insufficient success suggests that further investigation of allosteric regulation is essential to advance the drug development.

6. Aptamers

Aptamers, single-stranded deoxyribonucleic acid or ribonucleic acid oligonucleotides, are generated by an *in vitro* selection process called SELEX (systematic evolution of ligands by exponential enrichment). They can bind their target molecules with high specificity and selectivity, indicating the probable therapeutic and diagnostic applications for diseases like cancer, inflammatory diseases, etc. [70, 71]. Because aptamers contain some advantages over antibodies and other conventional small-molecule therapeutics, such as high specificity, flexible modification, and low adverse effect, they have been shown as a valuable substitute to protein antibodies [72]. Moreover, the strategies developed to chemically modify backbone can further improve affinity and bioavailability of aptamers [73]. Higher affinity and specificity could be simultaneously achieved by the genetic-algorithms-based ISM (in silico maturation) [74].

The properties above have paved the way to further studies on introduction of aptamers to preclinical and clinical applications. Based on previous data showing antitumor activity of AS1411, a first-in-class quadruplex DNA aptamer targeting nucleolus, a phase II trial found that AS1411 appears to have dramatic and durable responses in enrolled patients with metastatic renal cell carcinoma, even though about 34% participants have AS1411-related mild adverse events [75]. Malik *et al.* further discovered that AS1411-linked gold nanospheres (AS1411-GNS) could markedly promote superior cellular uptake by cancer cells and increase antiproliferative/cytotoxic effects, with no signs of toxicity [76]. Likewise, other clinical trials on aptmers targeting FIX (Coagulation Factor IX) [77], vWF (von Willebrand factor) [78], and TFPI (tissue factor pathway inhibitor) [79] respectively, all show that aptamers are well tolerated, safe, and represent a new promising target therapy. However, as for some side effects, further clinical investigations are warranted to better define the clinical indications, safety, efficacy, and optimal dosing strategy.

7. Decoys

Unlike antisense oligonucleotide approaches that target mRNA, decoys are short, double-stranded DNA molecules that compete with specific binding sites of transcription factors to prevent their binding at target promoters, in order to inhibit gene expression at pretranscription level. Since decoys are DNA, they are more stable and easy to handle than RNA-based intervention strategies [80]. Some methods, including the locked nucleic acid (LNA) introduced at the 3′-end [81] and chimeric decoys containing discrete binding sites [82], can increase decoys nuclease resistance and specificity. So far, numerous of studies have indicated that decoys are suited for novel potential therapeutic for combating cancer [80] and infectious diseases [83]. NOTCH1 decoy, a human IgG Fc consisting Notch1 extracellular domain inhibits tumor angiogenesis and growth by blocking Jagged-dependent activation of Notch signaling. Although well tolerated to mice for three weeks, NOTCH1 decoy treatment causes adverse severe gastrointestinal effects [84]. As above, the STAT3 (signal transducers and activators of transcription 3) decoy oligonucleotide represents another possible single-agent approach to targeting both the tumor and vascular compartments in murine tumor xenografts mediated through the inhibition of both STAT3 and STAT1 [85, 86]. Collectively, these findings point to decoys as highly attractive agents in gene-targeted therapy.

8. Concluding remarks

Gene-targeting strategies based on nucleic acid have opened a new era with the development of potent and effective gene intervention techniques, such as DNAzymes, ribozymes, siRNA, ASOs, aptamers, decoys, etc. It is demonstrated that these technologies have versatility and potency in disrupting pathophysiologically important pathways by silencing the target gene with relative specificity *in vivo* and *in vitro*. Numerous investigative works by several laboratories have been made in these fields. Although some clinical trials have proved the effectiveness of these techniques, only a few antisense drugs have been approved by the FDA for clinical purposes. The main difficulties on the way to develop successful nucleic acid drugs are as follows: how to ensure efficient and controlled delivery, prolonged target-specific action, and no adverse effects. If the challenges outlined above can be overcome, these molecules would prove to be valuable agents for economical and practical new therapies for diseases in the near future.

Author details

Zhijie Xu and Lifang Yang*

*Address all correspondence to: yanglifang99@hotmail.com

Cancer Research Institute, Central South University, Changsha, China

References

[1] Breaker RR, Joyce GF. A DNA enzyme that cleaves RNA. *Chem Biol* 1994;1:223-9.

[2] Fokina AA, Stetsenko DA, Francois JC. DNA enzymes as potential therapeutics: towards clinical application of 10-23 DNAzymes. *Expert Opin Biologic Ther* 2015;15:689-711.

[3] Santoro SW, Joyce GF. Mechanism and utility of an RNA-cleaving DNA enzyme. *Biochemistry* 1998;37:13330-42.

[4] Cairns MJ, King A, Sun LQ. Optimisation of the 10-23 DNAzyme-substrate pairing interactions enhanced RNA cleavage activity at purine-cytosine target sites. *Nucleic Acids Res* 2003;31:2883-9.

[5] Xu ZJ, Yang LF, Sun LQ, Cao Ya. Use of DNAzymes for cancer research and therapy. *Chin Sci Bull* 2012;57:3404-8.

[6] Kurreck J. Antisense technologies. Improvement through novel chemical modifications. *Eur J Biochem.* 2003;270:1628-44.

[7] Khan A, Benboubetra M, Sayyed PZ, Ng KW, Fox S, Beck G, et al. Sustained polymeric delivery of gene silencing antisense ODNs, siRNA, DNAzymes and ribozymes: in vitro and in vivo studies. *J Drug Target* 2004;12:393-404.

[8] Subramanian N, Kanwar JR, Akilandeswari B, Kanwar RK, Khetan V, Krishnakumar S. Chimeric nucleolin aptamer with survivin DNAzyme for cancer cell targeted delivery. *Chem Commun* 2015;51:6940-3.

[9] Li Y, Bhindi R, Deng ZJ, Morton SW, Hammond PT, Khachigian LM. Inhibition of vein graft stenosis with a c-jun targeting DNAzyme in a cationic liposomal formulation containing 1,2-dioleoyl-3-trimethylammonium propane (DOTAP)/1,2-dioleoyl-sn-glycero-3-phosphoethanolamine (DOPE). *Int J Cardiol* 2013;168:3659-64.

[10] Marquardt K, Eicher AC, Dobler D, Mader U, Schmidts T, Renz H, et al. Development of a protective dermal drug delivery system for therapeutic DNAzymes. *Int J Pharma* 2015;479:150-8.

[11] Xie J, Zhang S, Hu Y, Li D, Cui J, Xue J, et al. Regulatory roles of c-jun in H5N1 influenza virus replication and host inflammation. *Biochimica et Biophysica Acta* 2014;1842:2479-88.

[12] Cai H, Santiago FS, Prado-Lourenco L, Wang B, Patrikakis M, Davenport MP, et al. DNAzyme targeting c-jun suppresses skin cancer growth. *Sci Transl Med* 2012;4:139ra82.

[13] Yang L, Liu L, Xu Z, Liao W, Feng D, Dong X, et al. EBV-LMP1 targeted DNAzyme enhances radiosensitivity by inhibiting tumor angiogenesis via the JNKs/HIF-1 pathway in nasopharyngeal carcinoma. *Oncotarget* 2015;6:5804-17.

[14] Yang L, Xu Z, Liu L, Luo X, Lu J, Sun L, et al. Targeting EBV-LMP1 DNAzyme enhances radiosensitivity of nasopharyngeal carcinoma cells by inhibiting telomerase activity. *Cancer Biol Ther* 2014;15:61-8.

[15] Yang L, Lu Z, Ma X, Cao Y, Sun LQ. A therapeutic approach to nasopharyngeal carcinomas by DNAzymes targeting EBV LMP-1 gene. *Molecules* 2010;15:6127-39.

[16] Zhang J, Guo C, Wang R, Huang L, Liang W, Liu R, et al. An Egr-1-specific DNA-zyme regulates Egr-1 and proliferating cell nuclear antigen expression in rat vascular smooth muscle cells. *Exper Ther Med* 2013;5:1371-4.

[17] Hallett MA, Dalal P, Sweatman TW, Pourmotabbed T. The distribution, clearance, and safety of an anti-MMP-9 DNAzyme in normal and MMTV-PyMT transgenic mice. *Nucleic Acid Ther* 2013;23:379-88.

[18] Hallett MA, Teng B, Hasegawa H, Schwab LP, Seagroves TN, Pourmotabbed T. Anti-matrix metalloproteinase-9 DNAzyme decreases tumor growth in the MMTV-PyMT mouse model of breast cancer. *Breast Cancer Res* 2013;15:R12.

[19] Zhang M, Drummen GP, Luo S. Anti-insulin-like growth factor-IIP3 DNAzymes inhibit cell proliferation and induce caspase-dependent apoptosis in human hepatocarcinoma cell lines. *Drug Des Dev Ther* 2013;7:1089-102.

[20] Wiktorska M, Sacewicz-Hofman I, Stasikowska-Kanicka O, Danilewicz M, Niewiarowska J. Distinct inhibitory efficiency of siRNAs and DNAzymes to beta1 integrin subunit in blocking tumor growth. *Acta Biochimica Polonica* 2013;60:77-82.

[21] Shen L, Zhou Q, Wang Y, Liao W, Chen Y, Xu Z, et al. Antiangiogenic and antitumoral effects mediated by a vascular endothelial growth factor receptor 1 (VEGFR-1)-targeted DNAzyme. *Mol Med* 2013;19:377-86.

[22] Wang X, Zhang L, Ding N, Yang X, Zhang J, He J, et al. Identification and characterization of DNAzymes targeting DNA methyltransferase I for suppressing bladder cancer proliferation. *Biochem Biophys Res Comm* 2015;461:329-33.

[23] Yu X, Yang L, Cairns MJ, Dass C, Saravolac E, Li X, et al. Chemosensitization of solid tumors by inhibition of Bcl-xL expression using DNAzyme. *Oncotarget* 2014;5:9039-48.

[24] Kim JE, Yoon S, Choi BR, Kim KP, Cho YH, Jung W, et al. Cleavage of BCR-ABL transcripts at the T315I point mutation by DNAzyme promotes apoptotic cell death in imatinib-resistant BCR-ABL leukemic cells. *Leukemia* 2013;27:1650-8.

[25] Lai WY, Chen CY, Yang SC, Wu JY, Chang CJ, Yang PC, et al. Overcoming EGFR T790M-based tyrosine kinase inhibitor resistance with an allele-specific DNAzyme. *Mol Ther Nucleic Acids* 2014;3:e150.

[26] Tan CY, Weier Q, Zhang Y, Cox AJ, Kelly DJ, Langham RG. Thioredoxin-interacting protein: a potential therapeutic target for treatment of progressive fibrosis in diabetic nephropathy. *Nephron* 2015;129:109-27.

[27] Cao Y, Yang L, Jiang W, Wang X, Liao W, Tan G, et al. Therapeutic evaluation of Epstein-Barr virus-encoded latent membrane protein-1 targeted DNAzyme for treating of nasopharyngeal carcinomas. *Mol Ther* 2014;22:371-7.

[28] Liao WH, Yang LF, Liu XY, Zhou GF, Jiang WZ, Hou BL, et al. DCE-MRI assessment of the effect of Epstein-Barr virus-encoded latent membrane protein-1 targeted DNA-

zyme on tumor vasculature in patients with nasopharyngeal carcinomas. *BMC Cancer* 2014;14:835.

[29] Cho EA, Moloney FJ, Cai H, Au-Yeung A, China C, Scolyer RA, et al. Safety and tolerability of an intratumorally injected DNAzyme, Dz13, in patients with nodular basal-cell carcinoma: a phase 1 first-in-human trial (DISCOVER). *Lancet* 2013;381:1835-43.

[30] Krug N, Hohlfeld JM, Kirsten AM, Kornmann O, Beeh KM, Kappeler D, et al. Allergen-induced asthmatic responses modified by a GATA3-specific DNAzyme. *New Eng J Med* 2015;372:1987-95.

[31] Homburg U, Renz H, Timmer W, Hohlfeld JM, Seitz F, Luer K, et al. Safety and tolerability of a novel inhaled GATA3 mRNA targeting DNAzyme in patients with T2-driven asthma. *J Aller Clin Immunol* 2015;136:797-800.

[32] Turowska A, Librizzi D, Baumgartl N, Kuhlmann J, Dicke T, Merkel O, et al. Biodistribution of the GATA-3-specific DNAzyme hgd40 after inhalative exposure in mice, rats and dogs. *Toxicol Appl Pharmacol* 2013;272:365-72.

[33] Fuhst R, Runge F, Buschmann J, Ernst H, Praechter C, Hansen T, et al. Toxicity profile of the GATA-3-specific DNAzyme hgd40 after inhalation exposure. *Pulm Pharmacol Ther* 2013;26:281-9.

[34] Farra R, Grassi M, Grassi G, Dapas B. Therapeutic potential of small interfering RNAs/micro interfering RNA in hepatocellular carcinoma. *World J Gastroenterol* 2015;21:8994-9001.

[35] Borna H, Imani S, Iman M, Azimzadeh Jamalkandi S. Therapeutic face of RNAi: in vivo challenges. *Expert Opin Biologic Ther* 2015;15:269-85.

[36] Sioud M. RNA interference: mechanisms, technical challenges, and therapeutic opportunities. *Meth Mol Biol* 2015;1218:1-15.

[37] Ozcan G, Ozpolat B, Coleman RL, Sood AK, Lopez-Berestein G. Preclinical and clinical development of siRNA-based therapeutics. *Adv Drug Delivery Rev* 2015;87:108-19.

[38] Wittrup A, Lieberman J. Knocking down disease: a progress report on siRNA therapeutics. *Nature Rev Genet* 2015;16:543-52.

[39] Chung HY, Gu M, Buehler E, MacDonald MR, Rice CM. Seed sequence-matched controls reveal limitations of small interfering RNA knockdown in functional and structural studies of hepatitis C virus NS5A-MOBKL1B interaction. *J Virol* 2014;88:11022-33.

[40] Naito Y, Ui-Tei K. Designing functional siRNA with reduced off-target effects. *Meth Mol Biol* 2013;942:57-68.

[41] Fujita Y, Kuwano K, Ochiya T. Development of small RNA delivery systems for lung cancer therapy. *Int J Mol Sci* 2015;16:5254-70.

[42] Esposito CL, Catuogno S, de Franciscis V. Aptamer-mediated selective delivery of short RNA therapeutics in cancer cells. *J RNAi Gene Silencing* 2014;10:500-6.

[43] Li H, Tsui TY, Ma W. Intracellular delivery of molecular cargo using cell-penetrating peptides and the combination strategies. *Int J Mol Sci* 2015;16:19518-36.

[44] Kim MJ, Park JS, Lee SJ, Jang J, Park JS, Back SH, et al. Notch1 targeting siRNA delivery nanoparticles for rheumatoid arthritis therapy. *J Controlled Rel.* 2015;216:140-8.

[45] Cao C, Yan C, Hu Z, Zhou S. Potential application of injectable chitosan hydrogel treated with siRNA in chronic rhinosinusitis therapy. *Mol Med Rep* 2015;12:6688-94.

[46] Cui C, Wang Y, Yang K, Wang Y, Yang J, Xi J, et al. Preparation and characterization of RGDS/nanodiamond as a vector for VEGF-siRNA delivery. *J Biomed Nanotechnol* 2015;11:70-80.

[47] Golan T, Khvalevsky EZ, Hubert A, Gabai RM, Hen N, Segal A, et al. RNAi therapy targeting KRAS in combination with chemotherapy for locally advanced pancreatic cancer patients. *Oncotarget* 2015;6:24560-70.

[48] Strumberg D, Schultheis B, Traugott U, Vank C, Santel A, Keil O, et al. Phase I clinical development of Atu027, a siRNA formulation targeting PKN3 in patients with advanced solid tumors. *Int J Clin Pharmacol Ther* 2012;50:76-8.

[49] Schultheis B, Strumberg D, Santel A, Vank C, Gebhardt F, Keil O, et al. First-in-human phase I study of the liposomal RNA interference therapeutic Atu027 in patients with advanced solid tumors. *J Clin Oncol* 2014;32:4141-8.

[50] Martinez T, Gonzalez MV, Roehl I, Wright N, Paneda C, Jimenez AI. In vitro and in vivo efficacy of SYL040012, a novel siRNA compound for treatment of glaucoma. *Mol Ther* 2014;22:81-91.

[51] Moreno-Montanes J, Sadaba B, Ruz V, Gomez-Guiu A, Zarranz J, Gonzalez MV, et al. Phase I clinical trial of SYL040012, a small interfering RNA targeting beta-adrenergic receptor 2, for lowering intraocular pressure. *Mol Ther* 2014;22:226-32.

[52] Castanotto D, Stein CA. Antisense oligonucleotides in cancer. *Curr Opin Oncol* 2014;26:584-9.

[53] Agarwala A, Jones P, Nambi V. The role of antisense oligonucleotide therapy in patients with familial hypercholesterolemia: risks, benefits, and management recommendations. *Curr Atherosclerosis Rep* 2015;17:467.

[54] Rigo F, Seth PP, Bennett CF. Antisense oligonucleotide-based therapies for diseases caused by pre-mRNA processing defects. *Adv Exper Med Biol* 2014;825:303-52.

[55] Frazier KS. Antisense oligonucleotide therapies: the promise and the challenges from a toxicologic pathologist's perspective. *Toxicol Pathol* 2015;43:78-89.

[56] McClorey G, Wood MJ. An overview of the clinical application of antisense oligonucleotides for RNA-targeting therapies. *Curr Opin Pharmacol* 2015;24:52-8.

[57] Erba HP, Sayar H, Juckett M, Lahn M, Andre V, Callies S, et al. Safety and pharmacokinetics of the antisense oligonucleotide (ASO) LY2181308 as a single-agent or in combination with idarubicin and cytarabine in patients with refractory or relapsed acute myeloid leukemia (AML). *Invest New Drugs* 2013;31:1023-34.

[58] Buller HR, Bethune C, Bhanot S, Gailani D, Monia BP, Raskob GE, et al. Factor XI antisense oligonucleotide for prevention of venous thrombosis. *New Eng J Med* 2015;372:232-40.

[59] Monteleone G, Neurath MF, Ardizzone S, Di Sabatino A, Fantini MC, Castiglione F, et al. Mongersen, an oral SMAD7 antisense oligonucleotide, and Crohn's disease. *New Eng J Med* 2015;372:1104-13.

[60] Doherty EA, Doudna JA. Ribozyme structures and mechanisms. *Annu Rev Biophys Biomol Struct* 2001;30:457-75.

[61] Scott WG, Horan LH, Martick M. The hammerhead ribozyme: structure, catalysis, and gene regulation. *Progr Mol Biol Transl Sci* 2013;120:1-23.

[62] Asif-Ullah M, Levesque M, Robichaud G, Perreault JP. Development of ribozyme-based gene-inactivations; the example of the hepatitis delta virus ribozyme. *Curr Gene Ther* 2007;7:205-16.

[63] Nakano S, Kitagawa Y, Miyoshi D, Sugimoto N. Effects of background anionic compounds on the activity of the hammerhead ribozyme in Mg(2+)-unsaturated solutions. *J Biologic IInorg Chem* 2015;20:1049-58.

[64] Rouge JL, Sita TL, Hao L, Kouri FM, Briley WE, Stegh AH, et al. Ribozyme-Spherical Nucleic Acids. *J Am Chem Soc* 2015;137:10528-31.

[65] Paudel BP, Rueda D. Molecular crowding accelerates ribozyme docking and catalysis. *J Am Chem Soc* 2014;136:16700-3.

[66] Scarborough RJ, Gatignol A. HIV and Ribozymes. *Adv Exper Med Biol* 2015;848:97-116.

[67] Kobayashi H, Eckhardt SG, Lockridge JA, Rothenberg ML, Sandler AB, O'Bryant CL, et al. Safety and pharmacokinetic study of RPI.4610 (ANGIOZYME), an anti-VEGFR-1 ribozyme, in combination with carboplatin and paclitaxel in patients with advanced solid tumors. *Cancer Chemother Pharmacol* 2005;56:329-36.

[68] Mitsuyasu RT, Merigan TC, Carr A, Zack JA, Winters MA, Workman C, et al. Phase 2 gene therapy trial of an anti-HIV ribozyme in autologous CD34+ cells. *Nature Med* 2009;15:285-92.

[69] Morrow PK, Murthy RK, Ensor JD, Gordon GS, Margolin KA, Elias AD, et al. An open-label, phase 2 trial of RPI.4610 (Angiozyme) in the treatment of metastatic breast cancer. *Cancer* 2012;118:4098-104.

[70] Kang KN, Lee YS. RNA aptamers: a review of recent trends and applications. *Adv Biochem Eng Biotechnol* 2013;131:153-69.

[71] Ni X, Castanares M, Mukherjee A, Lupold SE. Nucleic acid aptamers: clinical applications and promising new horizons. *Curr Med Chem* 2011;18:4206-14.

[72] Li W, Lan X. Aptamer oligonucleotides: novel potential therapeutic agents in autoimmune disease. *Nucleic Acid Thera* 2015;25:173-9.

[73] Sun H, Zu Y. A Highlight of recent advances in aptamer technology and its application. *Molecules* 2015;20:11959-80.

[74] Savory N, Takahashi Y, Tsukakoshi K, Hasegawa H, Takase M, Abe K, et al. Simultaneous improvement of specificity and affinity of aptamers against Streptococcus mutans by in silico maturation for biosensor development. *Biotechnol Bioengin* 2014;111:454-61.

[75] Rosenberg JE, Bambury RM, Van Allen EM, Drabkin HA, Lara PN, Jr., Harzstark AL, et al. A phase II trial of AS1411 (a novel nucleolin-targeted DNA aptamer) in metastatic renal cell carcinoma. *Invest New Drugs* 2014;32:178-87.

[76] Malik MT, O'Toole MG, Casson LK, Thomas SD, Bardi GT, Reyes-Reyes EM, et al. AS1411-conjugated gold nanospheres and their potential for breast cancer therapy. *Oncotarget* 2015;6:22270-81.

[77] Vavalle JP, Rusconi CP, Zelenkofske S, Wargin WA, Alexander JH, Becker RC. A phase 1 ascending dose study of a subcutaneously administered factor IXa inhibitor and its active control agent. *J Thromb Haemo* 2012;10:1303-11.

[78] Bae ON. Targeting von Willebrand factor as a novel anti-platelet therapy; application of ARC1779, an Anti-vWF aptamer, against thrombotic risk. *Arch PharmaRes* 2012;35:1693-9.

[79] Gorczyca ME, Nair SC, Jilma B, Priya S, Male C, Reitter S, et al. Inhibition of tissue factor pathway inhibitor by the aptamer BAX499 improves clotting of hemophilic blood and plasma. *J Throm Haemo* 2012;10:1581-90.

[80] Rad SM, Langroudi L, Kouhkan F, Yazdani L, Koupaee AN, Asgharpour S, et al. Transcription factor decoy: a pre-transcriptional approach for gene downregulation purpose in cancer. *Tumour Biol* 2015;36:4871-81.

[81] Cogoi S, Zorzet S, Rapozzi V, Geci I, Pedersen EB, Xodo LE. MAZ-binding G4-decoy with locked nucleic acid and twisted intercalating nucleic acid modifications suppresses KRAS in pancreatic cancer cells and delays tumor growth in mice. *Nucleic Acids Res* 2013;41:4049-64.

[82] Brown AJ, Mainwaring DO, Sweeney B, James DC. Block decoys: transcription-factor decoys designed for in vitro gene regulation studies. *Anal Biochem* 2013;443:205-10.

[83] Jain B, Jain A. Taming influenza virus: role of antisense technology. *Curr Mol Med* 2015;15:433-45.

[84] Kangsamaksin T, Murtomaki A, Kofler NM, Cuervo H, Chaudhri RA, Tattersall IW, et al. NOTCH decoys that selectively block DLL/NOTCH or JAG/NOTCH disrupt angiogenesis by unique mechanisms to inhibit tumor growth. *Cancer Disc* 2015;5:182-97.

[85] Klein JD, Sano D, Sen M, Myers JN, Grandis JR, Kim S. STAT3 oligonucleotide inhibits tumor angiogenesis in preclinical models of squamous cell carcinoma. *PloS One* 2014;9:e81819.

[86] Sen M, Thomas SM, Kim S, Yeh JI, Ferris RL, Johnson JT, et al. First-in-human trial of a STAT3 decoy oligonucleotide in head and neck tumors: implications for cancer therapy. *Cancer Disc* 2012;2:694-705.

Chapter 9

Temperature-Dependent Regulation of Bacterial Gene Expression by RNA Thermometers

Yahan Wei and Erin R. Murphy

Additional information is available at the end of the chapter

http://dx.doi.org/10.5772/61968

Abstract

RNA thermometers (RNATs) are cis-encoded regulatory elements that modulate translational efficiently in response to environmental temperature. Since their initial discovery, numerous RNATs have been identified and characterized, with the majority of currently known RNATs present in a wide variety of bacterial species. RNATs repress translation at relatively low temperatures by physically preventing binding of the ribosome to the regulated transcript by incorporating the Shine-Dalgarno sequences (and/or start codon) into an inhibitory structure. As the environmental temperature increases, the inhibitory structure within the RNAT is destabilized and the repression of translation initiation is gradually relieved. With the development of identification techniques, the rate at which RNATs are identified, and the understanding of the molecular mechanisms governing their regulator function, has grown exponentially. With the ever-increasing number of characterized RNATs, broad families of these regulators have now been identified. It has also become abundantly clear that RNATs influence several essential physiological processes. This chapter aims to summarize the current knowledge of bacterial RNATs, with special emphasis placed on the molecular mechanisms underlying RNAT function, experimental techniques used to identify and characterize RNATs, families of bacterial RNATs, as well as biological processes controlled by RNATs, and future directions of the field.

Keywords: RNA thermometer, ribo-regulator, gene regulation, heat shock response, virulence factors

1. Introduction

Whether it is within a host or within the non-host environment, bacteria experience frequent, and often extreme, changes within their immediate environment. In order to survive and thrive under different environmental conditions, bacteria have evolved various systems that function to sense changes in environmental conditions and mediate rapid adaptation in response to the specific change. One condition that varies between the different environments encountered by pathogenic and non-pathogenic bacteria alike is temperature. Environmental temperature has direct effects on several fundamental biological processes, including proper folding of proteins and optimum activity of enzymes. To counteract the potentially detrimental effects of altered temperature, bacteria have evolved several strategies to respond to changes in environmental temperature, including specific heat shock and cold shock responses. Moreover, for pathogenic bacteria, a change of environmental temperature is a critical cue that can indicate entry into the host and/or progression of the disease process within an infected host. In order to establish and progress an infection, bacteria not only need to efficiently adapt to changing environmental conditions but also need to precisely regulate the production of specific virulence factors — processes that are dependent on the ability of bacteria to sense specific changes in environmental conditions, including temperature.

One method of sensing alterations in environmental temperature is through changes in the secondary structure of RNA molecules. Double-stranded regions within a given RNA molecule tend to dissociate into single-stranded structures with an increase in environmental temperature. The temperature at which half of the population of a given double-stranded RNA molecule is in the single-stranded conformation is defined as the Tm, a feature that is commonly used as a measurement for the stability of a given structure within an RNA molecule [1]. Due to its propensity to change conformation, an RNA structure that has a relatively low Tm is more responsive to changes in environmental temperature, a feature that facilitates its potential to act as a molecular thermosensors [2].

It is well established that translational efficiency is affected by the secondary structure of an RNA transcript, particularly that of the region containing the ribosome-binding site and/or start codon [3]. It was not until 1989, however, that the first *cis*-encoded temperature-responsive RNA regulatory element was identified [4]. Since that time, the rate at which temperature-responsive *cis*-encoded regulatory RNA elements have been identified, and the concurrent understanding of how they function to control target gene expression has grown exponentially — a statement that is particularly true of temperature-sensing RNA regulatory elements in bacteria. Based on their innate responsiveness to changes in environmental temperature, regulatory RNA elements that function to modulate the translational efficiency for the transcript in which they are housed in response to alterations in temperature have been termed "RNA thermometers" (RNATs) [5]. Unlike metabolites-binding riboswitches, the activity of RNAT is not modulated by the absence or presence of a ligand [6,7]. The regulatory function of an RNAT relies solely on its innate chemical nature, which dictates the differential stability of a specific inhibitory structure at different environmental temperatures.

With the ever-increasing number of characterized RNATs, variability within this class of regulators is now coming to light. While the majority of RNATs are composed of sequences within the 5′ untranslated region (5′ UTR) of the regulated gene, some have now been shown to be composed, at least in part, of sequences within the coding region of the regulated transcript or by sequences within the coding region of a preceding gene within a polycistronic transcript [8,9]. In addition, the number of stem loops composing different RNATs varies, ranging from one in the simplest RNATs to five in the most complex RNATs [10,11]. Despite the variability among RNATs, they all share several basic fundamental features. Identifying

http://dx.doi.org/10.5772/61968

and understanding the functional contribution of features conserved among characterized RNATs, as well as those that vary among this class of regulators, has and will continue to inform the foundational knowledge of the biological functions and chemical nature of these ubiquitous regulators. This chapter focuses on bacterial RNATs and provides a comprehensive summary of the current state of knowledge of RNATs, with emphasis given to discussions of the molecular mechanism underlying RNAT function, experimental techniques used to identify and characterize RNATs, families of bacterial RNATs, as well as the biological processes controlled by RNATs, and future directions of the field.

2. Molecular mechanism underlying the regulatory function of RNA thermometers

The molecular mechanism underlying the regulatory activity of RNATs is exquisitely simple, mediated entirely by temperature-induced structural changes within a target mRNA molecule. The currently proposed model of the molecular mechanism underlying RNAT function is that of a zipper [5]. More specifically, at relatively low "non-permissive" temperatures, an inhibitory structure is formed within the RNAT, at least in part, by binding of Shine-Dalgarno (SD) sequences with upstream sequences within the regulated transcript. Once formed, the inhibitory structure functions to block translation initiation by physically preventing binding of the ribosome to the regulated transcript. With an increase of temperature to that within a permissive range, the base-pairs that stabilize the inhibitory structure within the RNAT gradually dissociate, the ribosome-binding site becomes increasingly exposed and translation proceeds (Figure 1) [7,12].

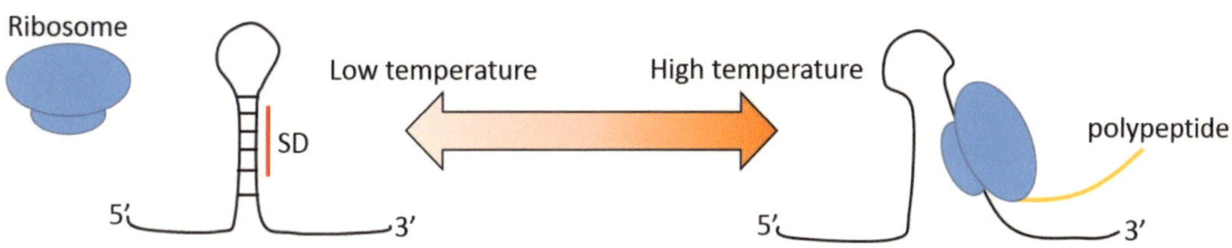

Figure 1. A schematic of the molecular mechanism underlying the regulatory function of RNA thermometers. In the figure, the hairpin structure indicates the inhibitory hairpin of an RNA thermometer, with the red line representing the region of Shine-Dalgarno (SD) sequence or ribosome-binding site. At relatively low, or non-permissive, temperatures, the formation of the inhibitory hairpin inhibits translation from the transcript by preventing binding of ribosome to the SD sequences. As the environmental temperature increases to the permissive range, the inhibitory structure dissociates, giving the ribosome access to the SD sequence and thus permitting translation.

Though responsive to temperature, RNAT-mediated regulation is not an all-or-nothing regulation but rather the shifting of an equilibrium towards an open or closed configuration depending on temperature [5]. Furthermore, mutagenesis-based experimentation has clearly demonstrated that it is the altered stability of the inhibitory structure rather than the primary sequence that plays the most critical role in the regulatory function of RNATs [13].

Several features differentiate RNATs from metabolites-binding riboswitches, a superficially related class of ribo-regulators. Firstly, as demonstrated by UV and NMR spectroscopy assays, the temperature-induced destabilization of the inhibitory structure within a given RNAT is a gradual and reversible process [7,14]. As a result of these fundamental features, RNATs mediate a graded response to temperature as opposed to an "on/off" type of regulation that is often associated with riboswitch-mediated regulation. Secondly, unlike riboswitches, structural changes within the RNAT are not mediated by an interaction with a small molecule or other cellular component, a foundational feature confirmed by *in vitro* structural analyses [7]. While the regulatory activity of RNATs is not responsive to the presence or absence of a ligand, Mg^{2+} has been found to facilitate the regulatory function of some RNATs [15]. Specifically, Mg^{2+} has been shown to affect the stability and thus the temperature responsiveness of the inhibitory structure of some RNATs, a feature that is fundamentally different than small molecule-induced switching between mutually exclusive structures as is seen with riboswitches.

While many advances have been made in recent years, several questions remain regarding the details of the molecular mechanism(s) underlying the activity of RNATs. For example, several studies investigating the regulatory mechanism of RNATs focus exclusively on the hairpin containing the SD sequence. As a consequence, the impact of additional structural features within an RNAT, particularly that of commonly observed upstream hairpins, remains largely unknown. Additionally, a recent study revealed that, for at least a subset of RNATs, the ribosome can bind to the SD sequence of the regulated transcript even at non-permissive temperatures when the inhibitory structure would be present [16]. Finally, while the current model of regulation invokes nothing more than temperature in mediating the structural changes that underlie the regulatory activity of RNATs, the role of additional factors, including that of the ribosome itself, remains the subject of active investigation.

3. Identification of RNA thermometers

3.1. *In silico* predictions

Given that the function of an RNAT is dependent on its structure, the identification of a new RNAT often starts with *in silico* analyses aimed at predicting secondary structure and the Gibbs free energy released by folding of known transcripts at different temperatures. Such predictions are often generated using freely available web-based programs, such as Mfold and are typically carried out individually for each transcript under investigation [17]. The identification of putative RNATs has been facilitated by the generation of a searchable database that contains the predicted structures within the untranslated regions of bacteria transcripts (RNA-SURIBA) [18]. Additionally, web servers such as RNAtips and RNAthermsw are now available, which calculate the folding energy of a given RNA molecule under varied temperatures [19,20]. Together, these databases and programs can be utilized to predict the presence of a putative RNAT within a given transcript or genome. While *in silico* approaches have proven powerful in the identification of many RNATs, they are limited in that they are only predictions. Additionally, all currently available *in silico* prediction tools are based on our current understanding of identified RNATs and therefore may not recognize novel types of RNATs with unique structural features. As the number of characterized RNATs continues to grow, so will our ability to accurately predict their presence in sequenced transcripts.

3.2. Identification with experimental approaches

Regardless of the approach used to predict the existence of a functional RNAT, the thermo-sensing regulatory activity of each putative element must be validated experimentally. There are several lab-based approaches currently being utilized to demonstrate the functionality of newly identified RNATs. One way in which the thermoresponsive regulatory activity of a putative RNAT is tested is to clone the element being investigated between a constitutive or an arabinose-inducible plasmid promoter and a reporter gene (e.g., *lacZ* or *gfp*) on a plasmid [10,11,21]. By introducing the constructed reporter plasmid into a bacterial strain and measuring the relative amounts of both the reporter transcript and reporter protein following growth of the strain at different temperatures, the functionality of the putative RNAT under investigation can be accessed. Specifically, if the putative RNAT is functional, production of the reporter protein will increase with the rise of temperature, while the relative levels of reporter transcript will not vary. Such experimental investigations, along with mutagenesis analysis, have been utilized to demonstrate the functionality of several newly identified RNATs [11,22–24] (Figure 2).

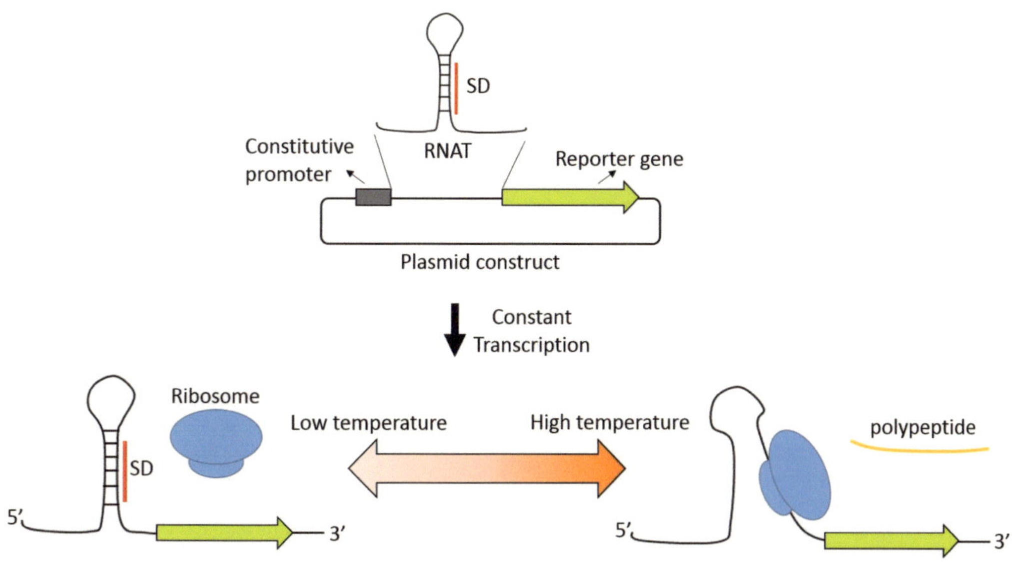

Figure 2. *In vivo* reporter plasmid-based assay used to experimentally test the thermoresponsive regulatory activity of a putative RNA thermometer. The reporter plasmid is constructed by cloning the inhibitory hairpin of a putative RNA thermometer between a constitutive or arabinose-induced plasmid promoter and a reporter gene. If the RNA thermometer is functional, it is expected that, following the introduction of the reporter plasmid into a bacterial strain and the growth of that strain at different temperatures, the relative amounts of the reporter transcript will be constant, while the relative levels of the reporter protein will be regulated in response to temperature.

To further validate the functionality of a predicted RNAT, *in vitro* analysis such as structure probing assays can be utilized to directly investigate the impact of varied temperature on the secondary structure of the element under investigation [22,25]. The principle underlying structure probing-based analyses is that specific RNA-digesting enzymes cleave RNA molecules based on the presence of specific secondary structures and/or primary sequences. For example, RNase T1 cleaves immediately 3′ to a single-stranded guanine, while RNase V1 cleaves double-stranded RNA in a sequence-independent manner. Briefly, to perform a structure probing analysis, the putative RNAT under investigation is synthesized by *in vitro* transcription and then radiolabeled at the 5′ end. Next, the labeled RNA molecule is subjected to partial digestion with various RNA-digesting enzymes separately, and the generated fragments visualized by electrophoresis in a denaturing polyacrylamide gel. By completing this analysis at different temperatures it is possible to determine the impact of environmental temperature on the global structure of the RNA molecule (Figure 3).

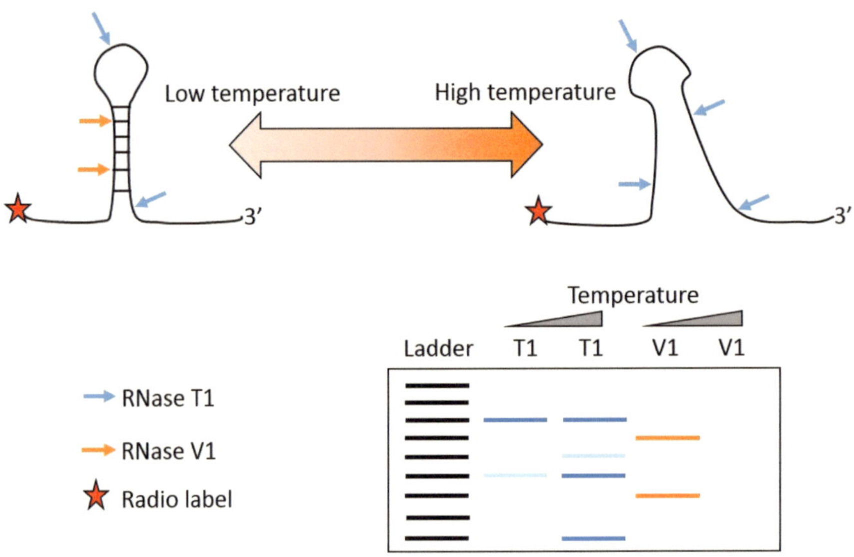

Figure 3. Structure probing assay used to experimentally determine the secondary structure of a putative RNA thermometer. An *in vitro* transcribed putative RNA thermometer is represented as the hairpin structure in this figure. Following radioactive end-labeling (indicated by a red star), the molecule is subject to partial digestions with different RNA-digesting enzymes and the resulting products are visualized by electrophoresis. In this figure, enzyme RNase T1 and RNase V1 are presented as examples of RNA-digesting enzymes, which cut, respectively, immediately 3′ to a single-stranded guanine and at double-stranded RNA in a sequence-independent manner. If the putative RNA thermometer under investigation changes conformation in response to alterations in temperature, an increase of environmental temperature would destabilize the inhibitory hairpin, resulting in a different pattern of radiolabeled fragments following digestion with the RNA degrading enzymes.

Moreover, techniques that study the physical properties of an RNA molecule, such as the nuclear magnetic resonance (NMR) spectroscopy and UV melting analysis, can be utilized to investigate the detailed base-pairing and their changes in response to temperature, thus revealing structural information as well as the molecular basis of thermosensing [7,26].

http://dx.doi.org/10.5772/61968

Together, these experimental analyses provide information about the dynamics of the inhibitory structure of a putative RNAT.

In addition to studies aimed at characterizing temperature-dependent changes in secondary structure, the regulatory activity of putative RNATs can be verified using a toe-printing assay, an *in vitro* analysis designed to directly assess the ability a ribosome to assemble and bind to the SD sequence contained on a given RNA molecule [22,27]. In the case of a functional RNAT, it would be predicted that ribosomal binding occurs at permissive temperatures, when the inhibitory structure is absent, but does not occur at non-permissive temperatures, when the inhibitory structure is present. Briefly, a putative RNAT is synthesized by *in vitro* transcription and incubated at a given temperature with a mixture of ribosome subunits and methionine conjugated tRNAs. If an SD sequence is available, a stable initiation complex will form on the RNA molecule, the presence of which is detected by reverse transcription using a radiolabeled primer that binds the RNA molecule downstream to the ribosome-binding site. The presence of the initiation complex will hinder the progression of the reverse transcriptase and thus result in the formation of a relatively short radiolabeled cDNA product. If the initiation complex cannot be formed, in this case because the SD sequences are occluded by the formation of an inhibitory structure within the putative RNAT, reverse transcription will not be blocked and a relatively long radiolabeled cDNA product will be generated. By completing toe-printing assays at different temperatures, the impact of temperature on the ability of the ribosome to interact with a putative RNAT can be directly determined. In the case of a functional RNAT, it would be expected that a relatively short cDNA product will be formed at permissive temperatures when the transcription initiation complex can form and that a relatively long cDNA product will be formed at non-permissive temperatures when assembling of the translation initiation complex is blocked by the formation of the inhibitory structures of the RNAT (Figure 4).

3.3. RNA structuromics

Recently, a combination of experimental and next-generation high-throughput techniques have been used to identify the structures of every RNA molecule within a single organism, collectively termed the "RNA structurome" [28]. Structuromic analyses performed at various temperatures have the potential to reveal a massive amount of information that will directly lead to the discovery of potentially expansive numbers of temperature-responsive regulatory RNA elements including RNATs [2]. The structurome of *Saccharomyces cerevisiae* and that of mice nuclear transcriptome were generated using parallel analysis of RNA structure (PARS) and fragmentation sequencing (Frag-seq), respectively [29,30]. The general experimental procedure that reveals the structurome of an organism includes two main steps: 1) structural probing of a certain transcriptome by specific RNA-digesting enzymes or chemicals that differentially cleave or modify RNA molecules based on the presence of specific secondary structures and 2) high-throughput sequencing analysis of the cDNA libraries generated from the digested/modified transcriptome. The structural probing portion of the analysis can be done either *in vitro* by treating the transcriptome harvested from the organism with structure- and sequence-specific RNA endonucleases or *in vivo* by cell-penetrating chemicals that modify

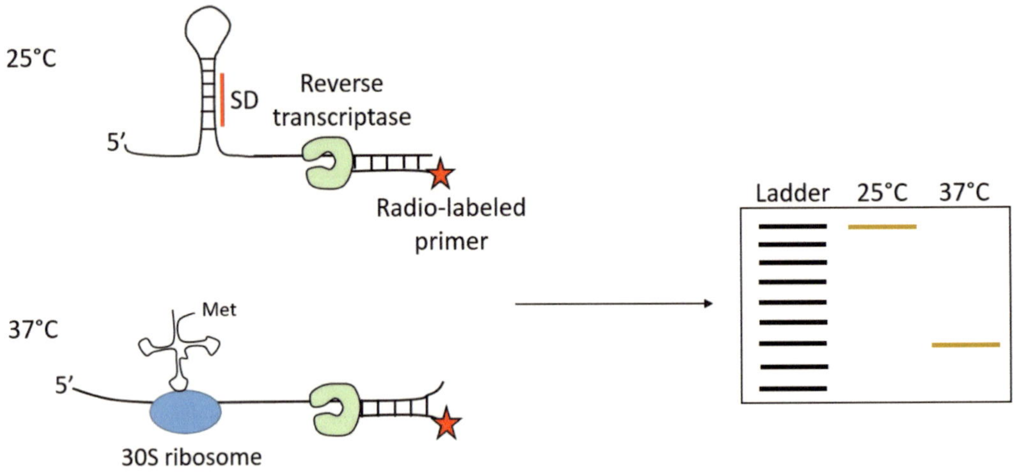

Figure 4. Toe-printing assay used to experimentally determine differential binding of the ribosome to a putative RNA thermometer at different temperatures. An *in vitro* transcribed putative RNA thermometer is incubated with a mixture of ribosome subunits and methionine-conjugated tRNA under different temperatures and then used as template in a reverse transcription reaction with a radiolabeled primer that binds downstream of the SD sequence. If the ribosome differentially binds the transcript, as would be predicted for that containing a functional RNA thermometer, reverse transcription would be expected to be hindered by the presence of the bound ribosome under permissive temperature (37°C in this example), thus producing a relatively short radiolabeled cDNA product. At non-permissive temperatures (25°C in this example), however, the ribosome would not be bound and reverse transcriptase would be expected to process to the end of the transcript generating a relatively long radiolabeled cDNA product.

or cleave single-stranded RNA bases within the cells [28,31]. This new experimental approach not only provides structural information of RNAs in physiological context but also evades the disadvantages of current *in silico* and *in vitro* analyses. Specifically, *in silico* prediction of RNA secondary structure is dependent on the length of RNA molecule that has been chosen; the longer the sequence, the lower the reliability of the prediction [32]. Additionally, secondary structures characterized by *in vitro* experimental analysis carry the caveat that the structures may be different *in vivo*. It is expected that, by completion of RNA structuromic analyses in a variety of bacterial organisms, the recognized numbers and types of RNATs will grow dramatically, an advancement that is critical to revealing the full impact of RNATs in controlling the physiology and virulence of bacterial species.

4. Families of RNA thermometers

The thermosensing activity of an RNAT is largely dependent on the physical features of its secondary structure, specifically by those features that impact the stability, or the Tm, of the inhibitory hairpin. In addition to the base-stacking interactions and the hydration shell of an RNA helix, other critical features of RNATs include 1) the number and stability of hairpins that are formed within the element; 2) the presence of canonical and non-canonical base-pairing within the inhibitory structure; 3) the existence of internal loops, bulges, or mismatches within the formed structure(s); and 4) the extent of base-pairing between sequences composing the SD site and/or start codon with upstream sequences contained on the transcript. Each of these features can directly impact the stability of the inhibitory structure within a given RNAT, which in turn dictates the responsiveness of the element to temperature. Despite sharing a common basic regulatory mechanism, differences in RNATs display different secondary structures and other key features, differences that are now used to classify bacterial RNATs into families. The two currently recognized families of RNATs are ROSE-like RNATs (repression of heat shock gene expression) and FourU RNATs. RNATs composing each of these two main families, as well as a few unique RNATs, are discussed below.

4.1. ROSE-like RNA thermometers

ROSE-like elements were first identified as conserved *cis*-regulatory elements located in the regions between the promoters and start codons of genes encoding small heat shock proteins (sHsps) in *Bradyrhizobium japonicum*. and within a short time were reported in other *Rhizobium* species as well as in *Agrobacterium tumefaciens* [21,33–35]. The heat shock response is a highly conserved process among microorganisms, and while their numbers vary between organisms, small heat shock proteins play a critical role in preventing protein denaturation and aggregation under heat stress. Based on the conservation of the biological process as well as the conservation of the primary sequence and secondary structure of the 17 originally identified ROSE-like elements, bioinformatics-based techniques were used to predict ROSE-like elements in the 5′ UTRs of sHsp encoding genes from 120 different archaea and bacteria [21,36]. As a result of these studies, 27 additional ROSE-like elements were identified in 18 different α- and γ-proteobacteria species [36]. Likely as a result of the approaches used to identify them, nearly all ROSE-like elements identified to date control the production of factors involved in the heat shock response. However, as additional ROSE-like RNATs are identified and characterized, it is expected that the contribution of these regulatory elements will be expanded beyond the production of heat shock response and into other physiological processes. This notion is supported by the recent identification of a ROSE-like RNAT in *Pseudomonas aeruginosa* that controls the production of rhamnolipids, a virulence factor that functions to protect the pathogen against killing by the human immune system [37]. Only with additional studies will the potentially expansive role of ROSE-like thermometers in controlling the physiology and virulence of bacterial species be revealed.

The ROSE-like family is the most extensively studied family of RNATs, harboring approximately 70% of all RNATs identified to date. All RNATs within the ROSE-like family are housed with 5′ UTR regions that range from 60 nucleotides to more than 100 nucleotides in length and that form 2 to 4 hairpins [36,37]. Within these hairpins, the 5′-proximal hairpin generally acts to stabilize the secondary structure and facilitate the correct folding of the other hairpins, while the 3′-proximal hairpin contains the SD region of the regulated transcript [7]. The defining features of ROSE-like RNATs that contribute to their temperature-responsive regulatory function include 1) the presence of a conserved anti-SD sequence 5′-UYGCU-3′ (Y stands for a pyrimidine) in the 3′-proximal hairpin, and 2) a "bulged" guanine within the SD sequestering hairpin (Figure 5) [36]. As a feature shared by all ROSE-like elements, it has been proposed that the "bulged" guanine within the SD sequestering hairpin is essential for the thermoresponsiveness of the regulatory element, a prediction that is supported by various mutagenesis-based experimental approaches and by NMR spectroscopy [7,38,39]. These studies have not only demonstrated that the "budged" guanine is essential for function but also revealed that the "bulged" guanine forms hydrogen bonds with the second guanine within the SD sequence of 5′-AGGA-3′. Additionally, towards the 3′ end of the SD site, two pyrimidines from the anti-SD strand form a triple-base pair with a uracil from the SD site with hydrogen bonds (Figure 5). The existence of two highly unstable pairs — a G-G pair and a triple-base pair — within the inhibitory hairpin of ROSE-like RNATs enables it to respond to the subtle changes of environmental temperature and thus to function as a temperature-sensitive regulatory element [7].

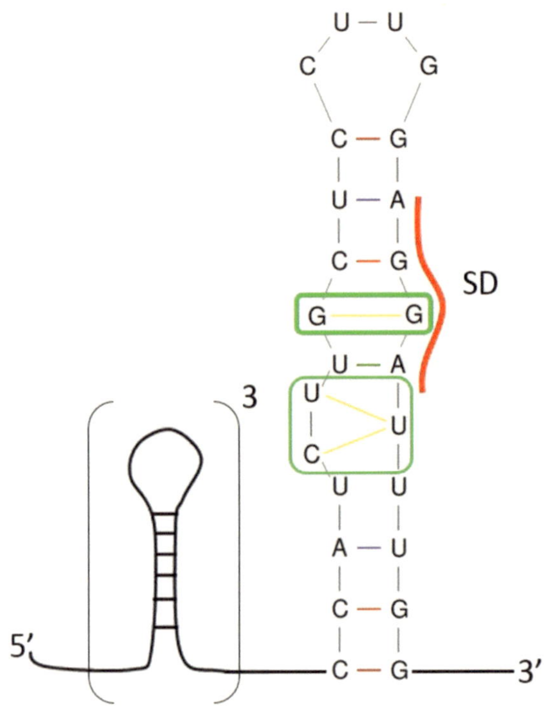

Figure 5. Structural features of the ROSE-like family of RNA thermometers, demonstrated by a schematic of the *hspA* RNA thermometer of *B. japonicum*. Within the four hairpins of *hspA* RNA thermometer, only the conserved structural features in the 3′ proximal hairpin are shown in detail, with the varied number of upstream hairpins indicated by the general hairpin structure in parentheses. The red line indicates the location of the SD sequence, while the conserved G-G pairing and triple-base pair are highlighted by the green boxes.

4.2. FourU RNA thermometers

FourU RNATs, so named due to the presence of four consecutive uracil residues within the SD sequestering inhibitory hairpin, represent the second family of currently identified RNATs. First identified in *Salmonella enterica,* a total of eight FourU RNATs have now been identified and characterized in a variety of bacterial species [10,11,22,40–42]. Unlike ROSE-like RNATs, only two characterized FourU RNATs function to control the production of a heat shock-related factor [10,22]. Instead, the majority of characterized FourU RNATs (*toxT* from *Vibrio cholera, lcrF/virF* from *Yersinia* species, as well as *shuA* from *S. dysenteriae* and its homologous gene *chuA* from some pathogenic *E. coli*) function to regulate the production of virulence factors in response to alterations in environmental temperature [11,40,41].

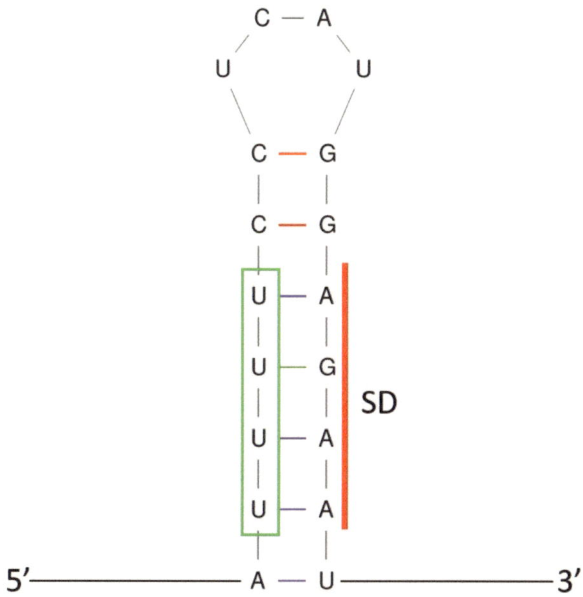

Figure 6. Structural features of the FourU family of RNA thermometers, demonstrated by a schematic of the *shuA* RNA thermometer of *S. dysenteriae*. Only the conserved portion of the inhibitory hairpin of *shuA* RNA thermometer is shown in this figure. The red line indicates the location of the SD sequences, while a green box indicates the location of the conserved four consecutive uracil residues.

The structural features of FourU RNATs are largely varied. For example, the length of the 5′ UTRs in which FourU RNATs are housed ranges from as short as 40 nucleotides (*htrA* from *E. coli* and *Salmonella*) to more than 280 nucleotides (*shuA* from *Shigella dysenteriae*) in length [10,11]. Additionally, the number of hairpins varies from a single hairpin with internal loops (*toxT* from *Vibrio cholerae*) to five hairpins, including an inhibitory hairpin with no internal loops (*shuA* from *S. dysenteriae*) [11,40]. Despite these differences, there are also several key features shared within FourU RNATs. The first shared feature is the presence of four consecutive uridine residues that form canonical A-U and/or non-canonical G-U base-pairs with SD sequences on the regulated transcript (Figure 6). Additionally, for RNATs within the FourU family, the SD sequestering hairpin is generated by no less than 5 continuous base-pairs, and often displaying conserved destabilizing features including the presence of relatively few G-C pairs, as well as internal mismatches or loops within the inhibitory structures. Likely due to the innate stability of the inhibitory hairpin within FourU RNATs, features destabilizing the inhibitory structure increase the responsiveness of FourU RNATs to temperature alterations, and disruption of these features result in altered thermosensing abilities. In the studies of each characterized FourU RNAT, mutagenesis analyses that introduce G-C base-pairs into the inhibitory structure result in the expected stabilization and, importantly, loss of thermosensing activity by the regulatory element [11,22,40]. NMR spectroscopy analysis has been utilized to study the dynamics of the inhibitory hairpin within the *agsA* FourU RNAT [15]. Specifically, a point mutation that introduces a C-G base-pair at the previously mismatched position adjacent to the SD region increased the melting temperature of the hairpin by 11°C. Additionally, two Mg^{2+} binding sites were found in the *agsA* FourU thermometer hairpin and it was demonstrated that Mg^{2+} functions to stabilize the inhibitory structure [15]. The degree to which these important features are conserved among members of the FourU RNAT family will be revealed only after additional members are identified and experimentally characterized. Such experimentation will not only define the FourU RNAT family of regulators but will also advance our ability to identify new FourU RNATs.

4.3. Additional types of RNA thermometers

It is important to note that not all characterized RNATs fit neatly into one of the two main families: ROSE-like and FourU. While all RNATs are thought to share a basic zipper-like thermosensing mechanism, several identified RNATs differ from those composing the main families in critical features, including primary sequence and/or secondary structure, features that impact the regulatory activity of these elements. It is the identification and characterization of the details of the molecular mechanisms underlying each of these additional types of RNATs that will expand our understanding of foundational principles governing RNA-mediated thermosensing.

In some RNATs, base-pairing involving the SD sequence is not complete but instead is disrupted by mismatches or "bulged" nucleotides, a feature also noted for ROSE-like elements. For example, the inhibitory structure within the RNATs that control the production of two putative lipoproteins LigA and LigB in *Leptospira interrogans* have identical nucleic acid sequences that include a mismatch of an adenine and a guanine within the SD sequestering hairpin [43]. Genes *hspX* and *hspY* that encode sHsps in *Pseudomonas putida* are also regulated by RNATs that contain one or two A·G mismatches disrupting the otherwise continued base-pairing of the SD region [8]. For these RNATs, further investigation is needed to understand the direct impact of the apparently conserved feature of mismatched or bulged sequences within the inhibitory structure on the regulatory activity of these elements.

Although lacking the presence of four consecutive uracil residues, two RNATs are similar to FourU RNATs in that they display more than 5 continuous base-pairs within the SD region of their inhibitory hairpins: one RNAT controls the production of an sHsp (Hsp17) from *Synechocystis* sp. PCC 6803, while the other controls the production of *Salmonella* GroES, a component of protein chaperon machinery [23,44]. RNAT-mediated regulation of *hsp17* is important for the survival of *Synechocystis* under heat stress, because Hsp17 not only prevents denatured proteins from aggregation but also protects the integrity of cellular membranes [45,46]. The 5′ UTR of *hsp17* has a single hairpin with an internal asymmetric loop [23]. In the SD sequence-binding region, instead of four uracils as seen in the FourU thermometer, the *hsp17* RNAT has a sequence of 5′-UCCU-3′ that forms four canonical pairs with the SD

http://dx.doi.org/10.5772/61968

sequence, including two G-C pairs. The remaining base-pairs in the inhibitory hairpin are mainly A-U pairs with two non-canonical G-U pairs. As the most stabled base-pairs within the *hsp17* RNAT, these two G-C base-pairs contribute to the stability and thus the inhibitory function of the hairpin. Other features such as the asymmetric internal loop and low ratio of G-C base-pairs destabilize the inhibitory structure, features that together enable the hairpin to dissociate with the increase of temperature. For the inhibitory hairpin of the *groES* RNAT, it has a mismatch of an adenine and a guanine that destabilizes this structure. While the secondary structure and temperature-responsive regulatory function of the *groES* RNAT has only been experimentally characterized in *Salmonella* and *E. coli*, this RNAT and its regulated factor, a necessary chaperon for proper folding of cellular components, are well conserved in enterobacteria [44].

For some RNATs, the function and stability of the inhibitory hairpin are impacted by base-pairing with sequences other than those within the SD region. For example, in the 5′ UTR of *prfA* from *Listeria monocytogenes*, a major portion of the SD region and the start codon are confined within internal loops and thus are partially single-stranded [47]. It has been demonstrated, however, that the hairpin within the *prfA* 5′ UTR containing the SD region and start codon does function as an RNA thermometer, an activity that is dependent on base-pairs that are located upstream of the SD site, which function to stabilize the unusually long hairpin. Another example of sequences other than those within the SD region that directly impact the regulatory function of an RNAT is the repeated nucleotide sequence of 5′-UAUACUUA-3′ in the RNAT of *cssA* from *Neisseria meningitides* [24]. These 8-nucleotide sequences are located upstream of the SD region and enable the RNAT to sense mild changes of environmental temperature, which is important for the survival of *N. meningitides.* As an opportunistic pathogen that colonizes only humans, it is important that *N. meningitidis* can sense and respond to a mild increase of temperature, as would be encountered during a fever response.

A unique example among currently identified RNATs is the one that controls the expression of *rpoH* in *E. coli* [9]. Binding of the ribosome to the SD region within the *ropH* transcript is facilitated by a sequence (named downstream box) located between the SD site and the start codon [48]. The *rpoH* RNAT inhibits translation via embedding this downstream box in the junction region of three stem loops instead of forming base-pairs within a single inhibitory hairpin as is the usual conformation in RNATs [9]. As the environmental temperature increases, two stems that paired with the downstream box melt at the junction position exposing the downstream box as a single strand, a conformation that facilitates ribosome binding to the transcript.

Lastly, there are currently three characterized RNATs that are located within intergenic regions of a polycistronic transcripts: *ibpB* from *E. coli, lcrF* from *Yersinia* species, and *hspY* from *P. putida* [8,41,49]. Their location within polycistronic transcripts differentiates these three RNATs from all others found in the 5′ UTR of monocistronic or polycistronic transcripts.

Although they display key features that differ from those possessed by RNATs in the ROSE-like or FourU families, many of the unique RNATs highlighted above are conserved between several bacterial species. There is little doubt that as additional bacterial RNATs are identified and characterized, commonalities will emerge and additional families will be recognized.

5. Bacterial processes controlled by RNA thermometers

The regulation of gene expression in response to changes in environmental temperature is important for survival of all bacteria and for virulence of pathogenic bacteria. RNATs have been found to confer efficient temperature-dependent regulation onto the expression of bacterial genes encoding factors involved in two critically important bacterial processes —heat shock response and virulence. In the following section, each of these two critical biological processes will be briefly introduced and the role that RNATs play in facilitating them will be discussed.

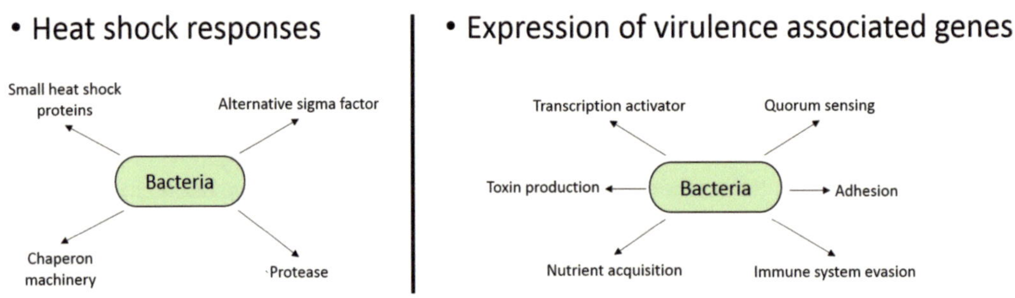

Figure 7. Key processes are controlled by RNA thermometers in bacteria. The influence of RNA thermometers on the bacterial heat shock response and on bacterial virulence is indicated by highlighting the different groups of genes whose expression is directly regulated by an RNA thermometer.

5.1. Heat shock response

The primary effect of increased temperature on bacteria is the resulting denaturation of temperature-sensitive proteins. Similar to regulatory RNA molecules, the function of a protein is strictly dependent on its structure, a feature that can be impacted by environmental temperature. Increased environmental temperature can result in partial or complete denaturation of a protein, resulting in a stable but often non-functional molecule [50]. In addition to the denaturation of proteins, high temperature is also associated with disruption of the bacterial cell membrane as well as damage to DNA molecules [46,51]. As a result of these effects, increased environmental temperature can be lethal to bacterial life and thus represents a stress that must be overcome.

In order to facilitate responsive adaptation to a rise in environmental temperature, bacteria express several genes that encode for factors that function to protect the organism from the detrimental effects generated by heat, collectively termed the heat shock response [52]. The main components of the heat shock response include 1) alternative sigma factors that direct the transcription of other heat shock responding genes; 2) heat shock proteins (Hsps), such as protein chaperon machinery that facilitate the proper folding of other proteins; 3) small Hsps (sHsps) that have multiple functions including preventing the formation of protein aggregates and protecting the integrity of cellular membrane; and 4) enzymes that degrade denatured proteins, repair damaged DNA, and more.

Understanding the molecular mechanisms underlying the temperature-dependent regulation of factors that facilitate the bacterial heat shock response is a major focus of ongoing investigations; the discovery of RNATs is rooted in these important studies. Since the identification of an RNAT that regulates the expression of a small heat shock protein (HspA from *B. japonicum*) and the heat shock alternative sigma factor σ^{32} (RpoH from *E. coli*), many other players in the heat shock response have been found to be regulated by RNATs, including chaperon component (GroES from *Salmonella*), heat-induced protease (HtrA from *E. coli*), and other small heat shock proteins [7,8,29,35] (Figure 7). Regulation of heat shock response is a complex process that involves the regulation of multiple factors at different steps of gene expression. That said, it seems that temperature-dependent regulation by RNAT is a fundamental regulatory mechanism that coordinately influences nearly all types of heat shock response factors. Given the high degree of conservation seen between heat shock responses factors produced by a wide variety of living organisms, temperature-dependent regulation mediated by RNATs is expected to be present in many organisms, including eukaryotic systems [52]. A finding that directly supports this prediction is that of a secondary structure within the 5′ UTR of *Drosophila* Hsps encoding mRNAs that functions to regulate translation from the transcript in response to environmental temperature [53]. The full extent of RNATs in controlling heat shock response in bacteria and beyond is yet to be revealed.

5.2. Virulence-associated genes of pathogenic bacteria

Once within the body of the host, and throughout the course of a natural infection, pathogenic bacteria face several challenges, including but not limited to 1) the need to adhere to host cells, 2) the need to evade killing by the host immune system, and 3) the need to acquire essential nutrients. To overcome these challenges and progress of an infection, bacteria produce specific virulence factors. As the production of virulence factors is most beneficial to an invading bacterium when it is within the host, several levels of regulation are often employed to ensure that the production of these important factors occurs only when the bacteria is within an environment that resembles that encountered within the infected host. RNATs are involved in regulating the production of a variety of virulence factors in several species of pathogenic bacteria, ensuring that these factors are most efficiently produced at the relatively high temperatures encountered within the infected host (Figure 7).

The expression of many virulence-associated genes is controlled by protein-based regulation, specifically that carried out by transcriptional regulators. Interestingly, RNATs have been found to directly control the production of three transcriptional activators that, in turn, function to control the expression of virulence-associated genes: *prfA* from *L. monocytogenes*, *lcrF* from *Y. pestis*, and *toxT* from *V. cholera* [40,41,47]. Another regulatory system that controls the expression of multiple virulence factors is quorum sensing. To date, one gene whose product is involved in quorum sensing-dependent modulation of virulence gene expression has been found to be regulated by an RNAT; this gene is *lasI* from *P. aeruginosa* [37]. RNATs within *lcrF* and *toxT* are FourU RNATs, while the RNATs controlling the expression of *prfA* and *lasI* have currently unique structure.

RNATs have also been implicated in controlling the expression of virulence-associated genes that encode factors involved in adhesion and immune evasion. For example, three virulence-associated genes in *N. meningitis* have been found to be regulated by RNATs: *cssA,* a gene encoding a factor involved in capsule production; *fHbp,* a gene encoding a factor H binding protein; and *lst,* a gene encoding a factor required for modifications of lipopolysacccharides [24]. In *L. interrogans, ligA* and *ligB,* two genes encoding putative lipoprotein, are also regulated by RNATs [43]. Additionally *P. aeruginosa rhlA,* a gene encoding an enzyme required for the synthesis of rhamnolipid, a compound that can prevent killing of the bacteria by host immune system, is regulated by an RNAT [37]. Except for *rhlA* RNAT, which is a member of the RSOE-like family, these other RNATs mentioned above have unique structures and thus are not members of the ROSE-like or FourU families of regulators.

To date, two genes involved in the acquisition of essential nutrients have been shown to be regulated by RNATs: *S. dysenteriae shuA,* a gene encoding an outer membrane heme-binding protein, and its homologous gene *chuA* in pathogenic *E. coli* [11]. Translation of *shuA/chuA* is controlled by a FourU RNAT located in the relative large 5′ UTR of the corresponding gene. Production of ShuA or ChuA facilitates the utilization of iron from heme, a potential source of essential iron found only within the relatively warm environment of the infected host [54].

For many pathogenic bacteria, the transmission from one host to the next involves exposure to different environments with different temperatures. The expression of many virulence-associated genes is influenced by environmental temperature, a signal that varies between the host and non-host environments. With an increasing number of virulence-associated genes that are now known to be regulated by the activity of RNATs, it is possible that temperature-dependent regulation mediated by RNATs will emerge as one of the basic regulatory strategies utilized by pathogenic bacteria. The full and potentially expansive role that RNATs play in controlling virulence of pathogenic bacteria is yet to be revealed.

6. Future directions

Although RNA-dependent regulation of gene expression has been a topic of active investigation for decades, investigations of RNATs are much more recent, with less than 100 RNATs having been identified to date (Table 1). Of note, RNATs vary in key structural features and influence different essential physiological processes.

RNAT type	Organism	Gene	Function of the gene	Reference
ROSE-element	*Agrobacterium tumefaciens*	*hspAT1* & *hspAT2*	Small heat shock protein	Balsiger *et. al.* 2004 [33]
	Bartonella henselae	*ibpA2*	Small heat shock protein	Waldminghaus *et. al.* 2005 [36]

RNAT type	Organism	Gene	Function of the gene	Reference
	Bartonella quintana	*ibpA2*	Small heat shock protein	Waldminghaus *et. al.* 2005 [36]
	Bradyrhizobium japonicum	*hspA, hspB, hspD, hspE,& hspH*	Small heat shock protein	Narberhaus *et. al.* 1998 [35]; Münchbach *et. al.* 1999 [34]
	Bradyrhizobium sp. *(Parasponia)*	*hspAP, hspCP, hspDP,& hspEP*	Small heat shock protein	Nocker *et. al.* 2001 [21]
	Brucella suis	*ibpA* & *hspA*	Small heat shock protein	Waldminghaus *et. al.* 2005 [36]
	Caulobacter crescentus	CC2258 & CC3592	Small heat shock protein	Waldminghaus *et. al.* 2005 [36]
	Erwinia carotovora	*ibpA* & *ibpB*	Small heat shock protein	Waldminghaus *et. al.* 2005 [36]
	Escherichia coli	*ibpA* & *ibpB*	Small heat shock protein	Waldminghaus *et. al.* 2005 [36]; Waldminghaus *et. al.* 2009 [39]; Gaubig *et. al.* 2011 [49]
	Mesorhizobium loti	*mll2387, mll3033, mlr3192,& mll9627*	Small heat shock protein	Nocker *et. al.* 2001 [21]
	Pseudomonas aeruginosa	*ibpA*	Small heat shock protein	Waldminghaus *et. al.* 2005 [36]; Krajewski *et. al.* 2013 [38]
	Pseudomonas putida	*ibpA*	Small heat shock protein	Waldminghaus *et. al.* 2005 [36]; Krajewski et. al. 2013 [38]
	Pseudomonas syringae	*PSPT02170*	Small heat shock protein	Waldminghaus *et. al.* 2005 [36]
		ibpA	Small heat shock protein	Krajewski *et. al.* 2013 [38]
	Rhizobium sp. strain NGR234	*hspAN* & *hspCN*	Small heat shock protein	Nocker *et. al.* 2001 [21]
	Rhodopseudomonas palustris	*RPA0054* & *hspD*	Small heat shock protein	Waldminghaus *et. al.* 2005 [36]
	Salmonella typhimurium	*ibpA* & *ibpB*	Small heat shock protein	Waldminghaus *et. al.* 2005 [36]

RNAT type	Organism	Gene	Function of the gene	Reference
	Shewanella oneidensis	*ibpA*	Small heat shock protein	Waldminghaus *et. al.* 2005 [36]
	Shigella flexneri	*ibpA* & *ibpB*	Small heat shock protein	Waldminghaus *et. al.* 2005 [36]
	Sinorhizobium meliloti	*ibpA* & *b21295*	Small heat shock protein	Waldminghaus *et. al.* 2005 [36]
	Vibrio cholerae	*hspA*	Small heat shock protein	Waldminghaus *et. al.* 2005 [36]
	Vibrio parahaemolyticus	*hspA*	Small heat shock protein	Waldminghaus *et. al.* 2005 [36]
	Vibrio vulnificus	*hspA*	Small heat shock protein	Waldminghaus *et. al.* 2005 [36]
	Yersinia pestis	*ibpA* & *ibpB*	Small heat shock protein	Waldminghaus *et. al.* 2005 [36]
	Pseudomonas aeruginosa	*rhlA*	Enzymes involved in the production of biosurfactant rhamnolipids	Grosso-Becerra *et. al.* 2014 [37]
FourU element	*Escherichia coli*	*htrA*	Stress-responding periplasmic protease	Klinkert et. al. 2012 [10]
	Salmonella enterica	*agsA*	Small heat shock protein	Waldminghaus *et. al.* 2007 [22]
		htrAp3	Stress-responding periplasmic protease (transcribed from the 3rd promoter of the gene)	Klinkert et. al. 2012 [10]
	Escherichia coli (some strains)	*chuA*	Outer membrane heme-binding protein	Kouse *et. al.* 2013 [11]
	Shigella dysenteriae	*shuA*	Outer membrane heme-binding protein	Kouse *et. al.* 2013 [11]
	Vibrio cholerae	*toxT*	Transcriptional activator of virulence factors (including cholera toxin)	Weber et. al. 2014 [40]
	Yersinia pestis	*lcrF*	Transcriptional activator of multiple virulence genes	Böhme *et. al.* 2012 [41]; Hoe *et. al.* 1993 [42]
	Yersinia pseudotuberculosis	*virF (lcrF)*	Transcriptional activator of multiple virulence genes	Böhme *et. al.* 2012 [41]

http://dx.doi.org/10.5772/61968

RNAT type	Organism	Gene	Function of the gene	Reference
Additional types	*Escherichia coli*	*rpoH*	Heat shock alternative sigma factor σ^{32}	Morita *et. al.* 1999 [9]
	Pseudomonas putida	*hspX* & *hspY*	Putative small heat shock proteins (similar to hspA,B,C)	Krajewski et. al. 2014 [8]
	Salmonella typhimurium	*groES*	Component of protein chaperon machinery	Cimdins *et. al.* 2013 [44]
	Synechocystis sp. PCC 6803	*hsp17*	Small heat shock protein	Kortmann *et. al.* 2011 [23]
	Leptospira interrogans	*ligA* & *ligB*	Putative lipoproteins promote adhesion -virulence related	Matsunaga *et. al.* 2013 [43]
	Listeria monocytogenes	*prfA*	Transcription activator of virulence factors	Johansson *et. al.* 2002 [47]
	Neisseria meningitidis	*cssA*	Capsule biosynthesis	Loh *et. al.* 2013 [24]
		fHbp	Factor H binding protein	
		lst	Lipopolysaccharide modification	
	Pseudomonas aeruginosa	*lasI*	Quorum sensing –synthesis quorum sensing signal	Grosso-Becerra *et. al.* 2014 [37]

Table 1. Summary of currently identified RNA thermometers

Despite their differences, all currently characterized RNATs are thought to share the same basic zipper-like temperature-responsive molecular mechanism, based on which both experimental and therapeutic applications can be derived. For example, artificial RNATs that have only a single hairpin to perform the temperature-dependent inhibition of translation have now been designed [55]. These artificial RNATs can be used as genetic tools to manipulate target gene expression. In the aspect of applying knowledge of RNATs in developing therapeutics, it is conceivable that compounds can be developed that would specifically stabilize the inhibitory structure within a given RNAT, thus decreasing expression of this target gene. Utilizing such an approach to inhibit the production of an essential gene product or virulence factor could prevent or limit infections by a variety of pathogenic bacteria.

Future applications of RNATs as genetic tools and/or drug targets are dependent on an increased understanding of these ubiquitous regulatory elements. With the maturation and development of experimental techniques, we could identify additional RNATs and study the molecular mechanisms underlying their regulatory activity in even greater detail. Moreover, due to the fundamental roles of RNA in the biological world, there is a great potential that RNATs also exist in archaea and eukaryotes. Further investigation and characterization of the conserved features and mechanisms of RNATs along with an understanding of the function of their regulatory targets could provide insight into the complex evolution of gene regulation. With the rate at which advances have been made in the field of RNA-mediated regulation, and specifically within the study of RNATs, there is no doubt that these and other important findings will be revealed sooner than later.

7. Nomenclature

cDNA library: The collection of single-stranded DNA products generated by reverse transcription using total RNAs isolated from an organism as templates.

Gibbs free energy: The thermodynamic potential of a system at a certain temperature and pressure. In this chapter, Gibbs free energy indicates the stability of a certain hairpin structure.

Heat shock response: The coordinated production of several proteins and other essential cellular components by a cell that work together to facilitate survival when the cell is exposed to an environmental temperature that is higher than its ideal surviving temperature.

Hydration shell: A shell-like structure formed by water molecules surrounding a molecule.

Melting temperature (Tm): The temperature at which half of the double-stranded molecules within a population assume a single-stranded conformation.

Quorum sensing: The coordinated regulation of bacterial gene expression in response to a secreted signal molecule that indicates the population density. When the signal molecule reaches a threshold amount, the cascade of signal-induced regulation occurs.

Riboswitch: A *cis*-encoded regulatory RNA element that functions to modulate target gene expression via switching between two mutually different secondary structures. Conformational changes within a riboswitch are induced by binding to a metabolite or other small molecule at a specific ligand-binding region.

RNA thermometer: A *cis*-encoded RNA element that represses translation via incorporation of the ribosome-binding site within an inhibitory hairpin at non-permissive temperatures. With increased temperature, the inhibitory structures within an RNA thermometer is destabilized, the ribosome-binding site is exposed, and translation proceeds.

Shine-Dalgarno (SD) sequence: Also known as the ribosome-binding site, the Shine-Dalgarno sequence is a sequence on an mRNA molecule to which the ribosome binds. Binding of a ribosome to an SD sequence on a transcript is necessary for the initiation of translation.

Sigma factor: A protein factor that facilitates the sequence-specific binding between an RNA polymerase and specific promoter regions on the DNA. Sigma factors are necessary for transcription initiation.

Structurome: The collective determination of the secondary structure of each transcript present in a given organism or cell type.

Transcriptome: The total population of RNA molecules present in a given organism or cell type.

http://dx.doi.org/10.5772/61968

5′ Untranslated region (5′ UTR): The region of a protein-encoding transcript that is located upstream of the translation start site. The 5′ UTR thus does not containing amino acid-coding sequences but rather contains the ribosome-binding site and often houses regulatory elements, such as RNA thermometers.

Acknowledgements

The authors would like to acknowledge both the Ohio University, Ohio University Heritage College of Osteopathic Medicine, and the American Heart Association for funding their ongoing studies of bacterial RNA thermometers.

Author details

Yahan Wei[1] and Erin R. Murphy[2*]

*Address all correspondence to: murphye@ohio.edu

1 Department of Biological Science, Ohio University, Athens, Ohio, U.S.A

2 Department of Biomedical Sciences, Ohio University Heritage College of Osteopathic Medicine, Athens, Ohio, U.S.A

References

[1] Wan Y, Qu K, Ouyang Z, Kertesz M, Li J, Tibshirani R, et al. Genome-wide measurement of RNA folding energies. Mol Cell. 2012 Oct;48(2):169–81.

[2] Wan Y, Kertesz M, Spitale RC, Segal E, Chang HY. Understanding the transcriptome through RNA structure. Nat Rev Genet. 2011;12(9):641–55.

[3] Gold L. Posttranscriptional regulatory mechanisms in *Escherichia coli*. Annu Rev Biochem. 1988 Jun;57(1):199–233.

[4] Altuvia S, Kornitzer D, Teff D, Oppenheim AB. Alternative mRNA structures of the cIII gene of bacteriophage lambda determine the rate of its translation initiation. J Mol Biol. 1989 Nov;210(2):265–80.

[5] Kortmann J, Narberhaus F. Bacterial RNA thermometers: Molecular zippers and switches. Nat Rev Microbiol. 2012;10(4):255–65.

[6] Serganov A, Nudler E. A decade of riboswitches. Cell. 2013;152(1-2):17–24.

[7] Chowdhury S, Maris C, Allain FH-T, Narberhaus F. Molecular basis for temperature sensing by an RNA thermometer. EMBO J. 2006;25(11):2487–97.

[8] Krajewski SS, Joswig M, Nagel M, Narberhaus F. A tricistronic heat shock operon is important for stress tolerance of *Pseudomonas putida* and conserved in many environmental bacteria. Environ Microbiol. 2014;16(6):1835–53.

[9] Morita MT, Tanaka Y, Kodama TS, Kyogoku Y, Yanagi H, Yura T. Translational induction of heat shock transcription factor σ^{32}: Evidence for a built-in RNA thermosensor. Genes Dev. 1999;13(6):655–65.

[10] Klinkert B, Cimdins A, Gaubig LC, Roßmanith J, Aschke-Sonnenborn U, Narberhaus F. Thermogenetic tools to monitor temperature-dependent gene expression in bacteria. J Biotechnol. 2012;160(1-2):55–63.

[11] Kouse AB, Righetti F, Kortmann J, Narberhaus F, Murphy ER. RNA-mediated thermoregulation of iron-acquisition genes in *Shigella dysenteriae* and pathogenic *Escherichia coli*. PLoS One. 2013 Jan;8(5):e63781.

[12] Rinnenthal J, Klinkert B, Narberhaus F, Schwalbe H. Direct observation of the temperature-induced melting process of the *Salmonella* fourU RNA thermometer at base-pair resolution. Nucleic Acids Res. 2010;38(11):3834–47.

[13] Nocker A, Hausherr T, Balsiger S, Krstulovic NP, Hennecke H, Narberhaus F. A mRNA-based thermosensor controls expression of rhizobial heat shock genes. Nucleic Acids Res. 2001;29(23):4800–7.

[14] Chowdhury S, Ragaz C, Kreuger E, Narberhaus F. Temperature-controlled structural alterations of an RNA thermometer. J Biol Chem. 2003;278(48):47915–21.

[15] Rinnenthal J, Klinkert B, Narberhaus F, Schwalbe H. Modulation of the stability of the *Salmonella* fourU-type RNA thermometer. Nucleic Acids Res. 2011;39(18):8258–70.

[16] Narayan S, Kombrabail MH, Das S, Singh H, Chary KVR, Rao BJ, et al. Site-specific fluorescence dynamics in an RNA "thermometer" reveals the role of ribosome binding in its temperature-sensitive switch function. Nucleic Acids Res. 2014;43(1):493–503.

[17] Zuker M. Mfold web server for nucleic acid folding and hybridization prediction. Nucleic Acids Res. 2003;31(13):3406–15.

[18] Waldminghaus T, Gaubig LC, Narberhaus F. Genome-wide bioinformatic prediction and experimental evaluation of potential RNA thermometers. Mol Genet Genomics. 2007;278(5):555–64.

[19] Chursov A, Kopetzky SJ, Bocharov G, Frishman D, Shneider A. RNAtips: Analysis of temperature-induced changes of RNA secondary structure. Nucleic Acids Res. 2013 Jul;41(Web Server issue):W486–91.

[20] Churkin A, Avihoo A, Shapira M, Barash D. RNAthermsw: Direct temperature simulations for predicting the location of RNA thermometers. PLoS One. 2014;9(4):e94340.

[21] Nocker A, Krstulovic NP, Perret X, Narberhaus F. ROSE elements occur in disparate rhizobia and are functionally interchangeable between species. Arch Microbiol. 2001;176(1-2):44–51.

[22] Waldminghaus T, Heidrich N, Brantl S, Narberhaus F. FourU: A novel type of RNA thermometer in *Salmonella*. Mol Microbiol. 2007;65(2):413–24.

[23] Kortmann J, Sczodrok S, Rinnenthal J, Schwalbe H, Narberhaus F. Translation on demand by a simple RNA-based thermosensor. Nucleic Acids Res. 2011 Apr;39(7): 2855–68.

[24] Loh E, Kugelberg E, Tracy A, Zhang Q, Gollan B, Ewles H, et al. Temperature triggers immune evasion by *Neisseria meningitidis*. Nature. 2013;502(7470):237–40.

[25] Ehresmann C, Baudin F, Mougel M, Romby P, Ebel JP, Ehresmann B. Probing the structure of RNAs in solution. Nucleic Acids Res. 1987;15(22):9109–28.

[26] Fürtig B, Richter C, Wöhnert J, Schwalbe H. NMR Spectroscopy of RNA. ChemBioChem. 2003;4(10):936–62.

[27] Hartz D, McPheeters DS, Traut R, Gold L. Extension inhibition analysis of translation initiation complexes. Methods Enzymol. 1988;164:419–25.

[28] Westhof E, Romby P. The RNA structurome: High-throughput probing. Nat Methods. 2010. p. 965–7.

[29] Kertesz M, Wan Y, Mazor E, Rinn JL, Nutter RC, Chang HY, et al. Genome-wide measurement of RNA secondary structure in yeast. Nature. 2010;467(7311):103–7.

[30] Underwood JG, Uzilov A V, Katzman S, Onodera CS, Mainzer JE, Mathews DH, et al. FragSeq: Transcriptome-wide RNA structure probing using high-throughput sequencing. Nat Methods. 2010 Dec;7(12):995–1001.

[31] Ding Y, Tang Y, Kwok CK, Zhang Y, Bevilacqua PC, Assmann SM. *In vivo* genome-wide profiling of RNA secondary structure reveals novel regulatory features. Nature. 2013;505(7485):696–700.

[32] Righetti F, Narberhaus F. How to find RNA thermometers. Front Cell Infect Microbiol [Internet]. 2014;4(September):1–6.

[33] Balsiger S, Ragaz C, Baron C, Narberhaus F. Replicon-specific regulation of small heat shock genes in *Agrobacterium tumefaciens*. J Bacteriol. 2004;186(20):6824–9.

[34] Münchbach M, Nocker A, Narberhaus F. Multiple small heat shock proteins in rhizobia. J Bacteriol. 1999;181(1):83–90.

[35] Narberhaus F, Käser R, Nocker A, Hennecke H. A novel DNA element that controls bacterial heat shock gene expression. Mol Microbiol. 1998;28(2):315–23.

[36] Waldminghaus T, Fippinger A, Alfsmann J, Narberhaus F. RNA thermometers are common in α- and γ-proteobacteria. Biol Chem. 2005;386(12):1279–86.

[37] Grosso-Becerra MV, Croda-Garcia G, Merino E, Servin-Gonzalez L, Mojica-Espinosa R, Soberon-Chavez G. Regulation of *Pseudomonas aeruginosa* virulence factors by two novel RNA thermometers. Proc Natl Acad Sci. 2014;111(43):15562–7.

[38] Krajewski SS, Nagel M, Narberhaus F. Short ROSE-like RNA thermometers control IbpA synthesis in *Pseudomonas* species. PLoS One. 2013;8(5).

[39] Waldminghaus T, Gaubig LC, Klinkert B, Narberhaus F. The *Escherichia coli ibpA* thermometer is comprised of stable and unstable structural elements. RNA Biol. 2009;6(4):455–63.

[40] Weber GG, Kortmann J, Narberhaus F, Klose KE. RNA thermometer controls temperature-dependent virulence factor expression in *Vibrio cholerae*. Proc Natl Acad Sci U S A. 2014 Sep 16;111(39):1–6.

[41] Böhme K, Steinmann R, Kortmann J, Seekircher S, Heroven AK, Berger E, et al. Concerted actions of a thermo-labile regulator and a unique intergenic RNA thermosensor control *Yersinia virulence*. PLoS Pathog. 2012;8(2).

[42] Hoe NP, Goguen JD. Temperature sensing in *Yersinia pestis*: Translation of the LcrF activator protein is thermally regulated. J Bacteriol. 1993 Dec;175(24):7901–9.

[43] Matsunaga J, Schlax PJ, Haake D a. Role for *cis*-acting RNA sequences in the temperature-dependent expression of the multiadhesive lig proteins in *Leptospira interrogans*. J Bacteriol. 2013;195(22):5092–101.

[44] Cimdins A, Roßmanith J, Langklotz S, Bandow JE, Narberhaus F. Differential control of *Salmonella* heat shock operons by structured mRNAs. Mol Microbiol. 2013;89(4): 715–31.

[45] Lee S, Prochaska DJ, Fang F, Barnum SR. A 16.6-kilodalton protein in the *Cyanobacterium synechocystis* sp. PCC 6803 plays a role in the heat shock response. Curr Microbiol. 1998 Dec;37(6):403–7.

[46] Török Z, Goloubinoff P, Horváth I, Tsvetkova NM, Glatz A, Balogh G, et al. *Synechocystis* HSP17 is an amphitropic protein that stabilizes heat-stressed membranes and binds denatured proteins for subsequent chaperone-mediated refolding. Proc Natl Acad Sci U S A. 2001;98:3098–103.

[47] Johansson J, Mandin P, Renzoni A, Chiaruttini C, Springer M, Cossart P. An RNA thermosensor controls expression of virulence genes in *Listeria monocytogenes*. Cell. 2002;110(5):551–61.

[48] Nagai H, Yuzawa H, Yura T. Interplay of two *cis*-acting mRNA regions in translational control of σ^{32} synthesis during the heat shock response of *Escherichia coli*. Proc Natl Acad Sci U S A. 1991;88(23):10515–9.

[49] Gaubig LC, Waldminghaus T, Narberhaus F. Multiple layers of control govern expression of the *Escherichia coli ibpAB* heat-shock operon. Microbiology. 2011;157(1): 66–76.

[50] Hartl FU, Hayer-Hartl M. Converging concepts of protein folding *in vitro* and *in vivo*. Nat Struct Mol Biol. 2009;16(6):574–81.

[51] Lindahl T. Instability and decay of the primary structure of DNA. Nature. 1993;362(6422):709-15.

[52] Verghese J, Abrams J, Wang Y, Morano KA. Biology of the heat shock response and protein chaperones: Budding yeast (*Saccharomyces cerevisiae*) as a model system. Microbiol Mol Biol Rev. 2012;76(2):115–58.

[53] Ahmed R, Duncan RF. Translational regulation of Hsp90 mRNA: Aug-proximal 5′-untranslated region elements essential for preferential heat shock translation. J Biol Chem. 2004;279(48):49919–30.

[54] Wyckoff EE, Duncan D, Torres AG, Mills M, Maase K, Payne SM. Structure of the *Shigella dysenteriae* haem transport locus and its phylogenetic distribution in enteric bacteria. Mol Microbiol [Internet]. 1998 Jun;28(6):1139–52.

[55] Neupert J, Karcher D, Bock R. Design of simple synthetic RNA thermometers for temperature-controlled gene expression in Escherichia coli. Nucleic Acids Res. 2008;36(19):1–9.

Chapter 10

A Review on the Thermodynamics of Denaturation Transition of DNA Duplex Oligomers in the Context of Nearest-Neighbor Models

João C. O. Guerra

Additional information is available at the end of the chapter

http://dx.doi.org/10.5772/62574

Abstract

In this review, we show that additive physical properties of DNA double strands can be written in terms of eight (polymeric) irreducible parameters. This results in self-consistency relations constraining the 10 duplex dimer contributions. Studies of thermodynamic stability of duplex oligomers are feasible, adding extra degrees of freedom, and this is performed, initially, considering the influence of end parameters on the thermodynamic stability of oligomers. Hence, we connect a statistical mechanics approach to the nearest-neighbor (NN) approach in the framework of the two-state model. This provides one correlation between end effects and initiation phenomena. Because of that, inside the framework of the NN modeling, the role played by end effects could not be so well defined. Thus, we propose a new model that permits to provide the nucleation free energies. The power of this model is relating the nucleation free energy to the mean composition of the chain, permitting to obtain a good estimate for the free energy associated only to the Watson–Crick base pairings.

Keywords: Thermodynamics of DNA duplex oligomers, Initiation free energy, Nucleation free energy, Irreducible parameters for free energy

1. Introduction

Many DNA biotechnological applications, such as PCR or cDNA expression profiling, depend on thermodynamic parameters, which are sequence dependent. We could cite the strand melting temperature as an example of such thermodynamic parameters. In a general way, physical properties of DNA or RNA sequences can be calculated, in a very simple form, from algorithms in the context of nearest-neighbor (NN) models, whose core characteristic is providing linear representations for experimental measurements on nucleotide chains always in terms of pairwise (dimer) sequence contributions.

However, NN dimer parameters cannot be assigned from experiments by solving a set of simultaneous linear equations. This is known since the beginning of the development of these models in the context of polynucleotide thermodynamic studies [1]. In fact, when we consider intrinsic composition closure constraints, the number of degrees of freedom of the model is effectively reduced.

Dimer occurrence relations are well known, thus allowing for decomposition of sequence properties into dimer contributions. Many authors, because of that, have preferred to use dimers as fundamental units because they provide the most straightforward decomposition scheme [2–6]. Although the dimer set values fit easily into the theoretical NN model approximation, the dimer composition is overstated. In fact, the dimer set size, which is equal to 16 (in the case of a simple chain) and 10 (in the case of double chains) [2–7], is greater than the number of degrees of freedom of the problem. However, the extraction of dimer set contributions has remained an ill-posed problem. To accomplish this task further, ad hoc regularization hypothesis has been used so far. As a corollary, so-far-unknown constraints must also link the full dimer set properties in some hidden way to restore full set unity. Alternative approaches have considered decompositions into irreducible and hence smaller sets of short sequences or dimer combinations [8–11]. Comparison between different laboratory sets and physical interpretation of set values becomes a difficult task due to the arbitrariness of possible renderings. The extraction of simpler and more direct dimer contributions from such sets has remained an ill-posed problem with no unique solutions but still embraced by a large community of biochemists [2–6]. To adopt the dimer set formulation further, ad hoc regularization hypotheses have been taken by different authors, such as the singular value decomposition method [4, 12].

In this review, among other objectives, we present an approach to this problem based on the analysis of how the nucleotide intrinsic intermolecular symmetries contribute to the structure of NN sets, as proposed by Licinio and Guerra [13]. Therefore, to achieve that, initially, it is introduced to a general quantum mechanics statement, giving physical properties for a sequence of heterogeneous molecules treated as subsystems assuming any of a given complete set of molecular states. The four-nucleotide set has a corresponding four-state representation. At this point, a careful choice of the number of degrees of freedom is made in order to project the representation into a three-dimensional molecular class space. Luckily, the three independent molecular classes are readily associated to the main biochemical classification of nucleotides as comprising purine–pyrimidine, amino–keto, and strong–weak bases. The representation of the four-nucleotide set as a tetrahedron in the three-dimensional space is at the heart of the approach, as proposed by Licinio and Guerra [13]. This representation has been used to generate DNA walks for sequence composition analysis or display. The corresponding proper space metrics have also been recently used for phylogenetic sequence comparisons [14]. In the following, we proceed to contract the original quantum mechanics statement into an irreducible formulation using the four-nucleotide tetrahedron representation. This molecular symmetrical decomposition is found to provide the right number of fundamental properties (free parameters), which is equal to 8, for the case of DNA double strands. We shall refer to these fundamental properties as constituting a symmetrical set of irreducible tensorial parameters. Next, we relate this decomposition to the dimer set formulation. The comparison uncovers useful and so far hidden self-consistency relations among dimers.

However, an important point still would need to be clarified. In fact, in many publications, one finds datasets that include experimental values for duplex oligonucleotides, where end effects were believed to be important [2–6]. Nevertheless, such initiation and termination parameters would seem to be very sensitive to the modeling and have changed a lot even inside the same research group [3–6]. In fact, Xia et al. had already argued that data from melting experiments of RNA duplexes are of insufficient accuracy to distinguish end effects [15]. With this motivation, as a second step in the development of the approach proposed by us and presented in this review, we proposed to extend the irreducible model to investigate how it would accommodate end effects. Guerra and Licinio in fact performed such extension and calculated the irreducible parameters for free energy, entropy, enthalpy, and the respective end contributions [16]. Later, a detailed algorithm for performing such calculations is described. However, at this point, it is necessary to anticipate some conclusions. For example, Guerra and Licinio obtained values for the end effects with relatively large errors. In addition, specifically for free energy, they could not distinguish between the weak and strong terminal base pairs. In the light of their finding, one simple statistical mechanics approach, when applied to the melting transition, shows that the approach based on end effects, according to the NN approach, proves to be naive, even heuristic. In fact, since the end effects were initially (wrongly) identified as the nucleation free energies, they should be dependent on the mean global composition of the chain. However, an only slightly more detailed statistical mechanics approach can show that, summed to the eight (polymeric) irreducible parameters for free energy, as already mentioned, there are other two parameters related to the initiation of the double helix (related to two possible base pairings). That is, in the light of the NN approach, there are 10 parameters, which expand the free energy of any DNA oligomers [17].

Before we continue our discussion throughout the forthcoming sessions, it is important to inform the reader that all theoretical results we obtained were applied to the analysis of DNA free energy by introducing, initially, the formulation of end contributions to the model, which will be presented later in this chapter. A simple statistical mechanics approach is then applied to the problem. As a result, a second set of parameters, including this time the initiation parameters, will be obtained. Anyway, a self-consistent set has thus been fit to free energy data from 108 short duplex oligomer sequences as available in the literature. We will show that, using both the modeling, the first based on end effects and the second based on the use of double helix initiation parameters, the more compact and symmetrical self-consistent set is shown to provide at least as good modeling for oligomer free energy as standard NN dimer models. The far-reaching strength of the theoretical modeling frame for DNA or RNA sequences as proposed by us resides in its compactness and symmetry. As will be discussed later in this review, one of the immediate and practical consequences of the use of the tetrahedral model is the disclosure of the initially hidden dimer self-consistency relations.

2. A quantum mechanics formulation for sequence properties

Complexity in biological phenomena represents an enormous challenge and a rich field for the application and development of physical methods. To unfold simple biopolymer phenomena, we start by a biochemical meaningful nucleotide representation into molecular classes and count on tools provided by the quantum mechanics. Here, we shall use the quantum mechanics formulation based on the matrix representation. What is needed from start is some base set for the description of the states of the system, which, for us, is a DNA or RNA sequence. The ensemble of sequence states is given by allowable sequence composition alone. We want to describe and isolate gross composition states. Inner electronic states or molecular conformation contributions, which would require a much finer level of quantum description, are so far intrinsically averaged. State transitions are of course forbidden if one neglects mutations. The sequence state will be given in terms of its molecular constitution, and a nucleotide set representation will condition the sequence representation.

The quantum mechanics expectation for any observable is given in terms of the corresponding operator $\boldsymbol{\Theta}$ and system state $|\Psi\rangle$ as $\langle\psi|\boldsymbol{\Theta}|\psi\rangle$, in Dirac's notation. The state of a system comprising N particles or molecules is usually expressed as the tensorial product of their component states $|b(i)\rangle$, $(1 \leq i \leq N)$:

$$\left|\Psi\right\rangle = \left|b(1)\right\rangle\left|b(2)\right\rangle \otimes \cdots \otimes \left|b(N)\right\rangle = \left|b(1);b(2);\ldots;b(N)\right\rangle \tag{1}$$

For d-dimensional component states, this would lead a priori to the specification of $(Nd)^2$ operator matrix elements $\mu(i)\nu(j)$. If interaction range is limited, however, then many off-diagonal matrix elements become null, and a reduced formulation can be sought. Considering only sequential NN interactions, the expectation can thus be written simply as

$$\mathrm{E} = \sum_i \left\langle b(i);b(i+1)\right|\boldsymbol{\Theta}\left|b(i);b(i+1)\right\rangle \tag{2}$$

Here, submatrix elements pertaining to the same component at position i (diagonal or self-matrices $\boldsymbol{\Theta}_{\mu(i)\nu(j)}$), which are internal to the sequence ($i \neq 1, N$), should be halved because they are counted twice in this formulation (see Fig. 1). We hope further reduction of this development can be obtained considering implicit symmetries of the Hermitian $\boldsymbol{\Theta}$ matrix and its invariants under orthonormal base representations.

3. Nucleotide class-state representation

The most straightforward representation for a four-nucleotide set is, obviously, a four-dimensional vector. This "independent-nucleotide" representation has been implicitly

http://dx.doi.org/10.5772/62574

adopted by many authors and leads to 4 × 4 matrices or 16 parameter sets when considering nucleotide pairwise properties [11]. This representation, however, already overstates the nucleotide composition problem from the beginning. The set representation should be more concisely established in a three-dimensional space. Thus, a complete and symmetrical representation for the usual DNA (or RNA) four-nucleotide set can be given within a tetrahedral decomposition scheme into a three-dimensional orthonormal base set $|x\rangle$, $|y\rangle$, $|z\rangle$. The pure nucleotide states $|b(i)\rangle$ are given as follows [14]:

$$|A\rangle = \begin{pmatrix} 1 \\ 1 \\ 1 \end{pmatrix}; |T\rangle = \begin{pmatrix} -1 \\ -1 \\ 1 \end{pmatrix}; |C\rangle = \begin{pmatrix} -1 \\ 1 \\ -1 \end{pmatrix}; |G\rangle = \begin{pmatrix} 1 \\ -1 \\ -1 \end{pmatrix} \tag{3}$$

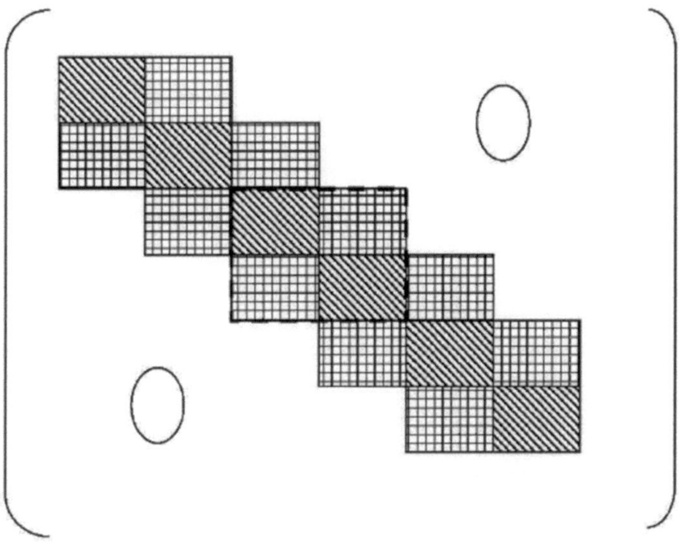

Figure 1. Structure of an expectation matrix for a sequence of $n = 6$ identical components (molecules in arbitrary states). The components have d degrees of freedom represented through d orthogonal base states, which result in $3n–2 = 16$ submatrices of size d^2. In this case, only nearest-neighbor interactions are considered. The matrix above corresponding to the quantum mechanics formulation of Eq. 1 is Hermitian and periodic, allowing for a more synthetic representation. One periodic module of four submatrices implicit in Eq. 2 has been distinguished by a dashed line. Observe that internal submatrices in the diagonal are counted twice according to the formulation of Eq. 2 [13].

The nucleotides themselves are represented as a nonorthogonal (tetrahedral) $\sqrt{3}$ -modulus vector set (Fig. 2). The four-nucleotide states are not independent and can be expressed in terms of three independent abstract nucleotide class states. Due to this decomposition, z-component discriminates weak (two bridges, AT) versus strong (three bridges, CG) hydrogen bonding for Watson–Crick (WC) pairing; x-component discriminates purine (double ring, AG) versus pyrimidine (single ring, CT) nucleotide sizes; and y-component discriminates amino (nitrogen containing, AC) versus keto (oxygen containing, GT) nucleotide radicals.

In quantum mechanics language, a $|x\rangle$ base state, for example, is a ring number or purine–pyrimidine class state, whereas $|A\rangle = |x\rangle + |y\rangle + |z\rangle$ is an adenine molecular state decomposed in terms of proper nucleotide class subspaces. Any pure nucleotide state can thus be represented in terms of molecular class states.

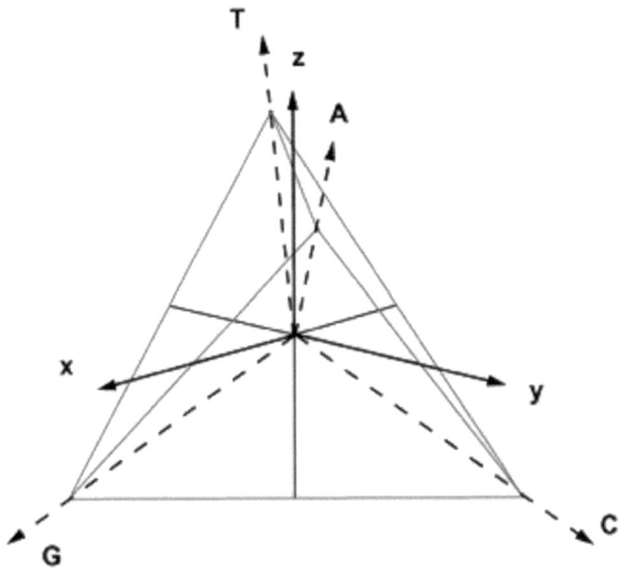

Figure 2. Orthonormal x–y–z base set and tetrahedral DNA-nucleotide set representation. Each of the three axes distinguishes a specific molecular class feature. Purines are distinguished from pyrimidines through x-coordinate. Amino is distinguished from keto through y-coordinate. And, finally, weak WC hydrogen-bridge binding is distinguished from stronger binding through z-coordinate [14].

Each possible nucleotide pair shares one of its fundamental molecular structural characteristics as a group in a given class, which differs from the complementary pair as another group in the same class. This is latent when we observe Eq. 3, which translates perfectly well the intrinsic cubic symmetry of the tetrahedron. From now, we proceed to construct our approach, which will use a complete nucleotide representation, and, then, having seen based on this representation, it will provide properties associated to each molecule decomposing them in terms of three differential affinity groups or classes. Therefore, the choice of a tetrahedral set is thus natural and convenient for its intrinsic orthogonality and symmetry properties, which are related to common molecular group classifications. Nevertheless, its main advantage is to fulfill the necessity for a three-dimensional bijective representation of a four-set composition.

4. Irreducible representation

Returning to the quantum mechanics formulation, our intention is to exploit remaining invariants and redundancies from the structure of the matrix operator present in Eq. 2 in order to further reduce its number of parameters. The three-dimensional nucleotide basis should be kept in mind. The sequence-dependent states of an observable will then assume discrete values given by a most compact expansion of its expectation as follows:

$$\mathbf{E}=\sum_{i}\left(S+V\left|b(i)\right\rangle+\left\langle b(i)\right|M\left|b(i+1)\right\rangle\right) \tag{4}$$

in substitution to Eq. 2; in Eq. 4, $|b(i)\rangle$ are still the sequence nucleotide states at coordinate i, which are given in terms of class states by Eq. 3.

The bracket notation indicates vector and dyadic contractions as usual. The expansion in Eq. 4 is quite intuitive, in the sense that the first two terms represent linear contributions to a property from the sequence composition, whereas the third term comprises nonlinear effects due to NN interference or differential stacking interactions. Comparison with Eq. 2 allows the identification of its components. The first term is a constant or mean contribution to the observable, given as the invariant trace of the square expectation periodic matrix $S=Tr(\ \boldsymbol{\Theta})$. The trace represents a molecular state independent contraction of the self-matrix diagonal, where, by construction, any pure nucleotide component (Eq. 3) equally squares to one ($b_\mu^2=1$). The remaining cross terms of the self-matrix similarly contract to a vector because all pure nucleotide states $|b(i)\rangle$ also have cyclically multiplicative class components ($b_x=\ b_y b_z$, etc.). This contraction gives the second term as an order-independent or global-composition contribution, with components $\langle V|=4\ Re\ (\boldsymbol{\Theta}_{y(1)z(1)},\ \boldsymbol{\Theta}_{x(1)z(1)},\ \boldsymbol{\Theta}_{x(1)y(1)})$. The third term is an NN or first-order sequence stacking contribution to the observable. The stacking matrix M is a second-rank tensor and has its elements given from the cross expectation matrix as $M_{\mu\nu}=2\ Re(\boldsymbol{\Theta}_{\mu(1)\nu(2)})$. The symmetrical sum of the expectation matrix Hermitian conjugates results in a fully contracted real formulation.

Decomposition of nucleotide sequence observable expectation as given in Eq. 4 naturally leads to an irreducible 13-parameter description of physical properties (S, V_μ, and $M_{\mu\nu}$), which we call the symmetrical set, within the NN approximation. Note that a traditional description of stacking-dependent properties is often stated in terms of the NN dimer composition, that is, as a linear combination of the 16-ordered 5′-3′ NN dimer set E_{ij} :

$$\mathbf{E}=\sum_{i,j=A,\,T,\,C,\,G} N_{ij}\mathbf{E}_{ij} \tag{5}$$

However, the NN dimer set is overspecified, that is, only a smaller set of NN combinations can be a priori obtained from inversions of Eq. 5 because Eq. 5 is supplemented by independent composition closure relations. For implicit circular sequences (or for very long sequences, i.e., polynucleotides), these can be taken as any three of the following:

$$\begin{aligned}
&\sum_{b=A,T,C,G}\left(N_{Ab}-N_{bA}\right)=0\\
&\sum_{b=A,T,C,G}\left(N_{Tb}-N_{bT}\right)=0\\
&\sum_{b=A,T,C,G}\left(N_{Cb}-N_{bC}\right)=0\\
&\sum_{b=A,T,C,G}\left(N_{Gb}-N_{bG}\right)=0,
\end{aligned} \tag{6}$$

reducing the number of independent dimers in the set to arbitrary 13. Similar arguments hold for linear oligomers. In comparison, the decomposition of physical properties in the symmetrical set proposed here is in a fundamental level; since from the beginning, it includes only a priori linearly independent terms and gives contributions to the observable in the hierarchic form of three expectation tensors of increasing rank, corresponding to different levels of analysis. The 16-NN expectations can otherwise be easily obtained as a linear combination of the 13 symmetrical-set tensor components. In that case, it is useful to rewrite Eq. 4 in a form appropriate for NN dimer decomposition as follows:

$$\mathbf{E}_{b(1)b(2)} = S + \left\langle V\middle|\frac{b(1)+b(2)}{2}\right\rangle + \left\langle b(1)\middle|M\middle|b(2)\right\rangle, \tag{7}$$

where, to correctly account for additivity, as given by Eq. 5 for each dimer in a sequence, the two nucleotide linear contributions are halved. Explicitly, one has applying Eq. 3 to Eq. 7:

$$\begin{aligned}
\mathbf{E}_{TA} &= S+V_z-M_{xx}-M_{xy}-M_{xz}-M_{yx}-M_{yy}-M_{yz}+M_{zx}+M_{zy}+M_{zz}\\
\mathbf{E}_{AT} &= S+V_z-M_{xx}-M_{xy}+M_{xz}-M_{yx}-M_{yy}+M_{yz}-M_{zx}-M_{zy}+M_{zz}\\
\mathbf{E}_{CA} &= S+V_y-M_{xx}-M_{xy}-M_{xz}+M_{yx}+M_{yy}+M_{yz}-M_{zx}-M_{zy}-M_{zz}\\
\mathbf{E}_{TG} &= S-V_y-M_{xx}+M_{xy}+M_{xz}-M_{yx}+M_{yy}+M_{yz}+M_{zx}-M_{zy}-M_{zz}
\end{aligned} \tag{8}$$

and so on. Tensor elements can be either conversely determined from reported dimer values or self-consistently derived from fits to raw polynucleotide data using Eqs. 8 and 5, or directly from Eq. 4, while from a theoretical point of view, molecular symmetry arguments or ab initio calculations could be used to guess tensor structure and values.

4.1. Double strands

For measurements concerning double strands, aside end effects, it is well known that complementary strand symmetry further reduces the problem to the statement of only 10 conjugated NN dimer pair values (see the expressions in Eq. 12) linked through two independent composition closure relations as follows:

http://dx.doi.org/10.5772/62574

$$\sum_{b=A,T,C,G} \left(N_{Ab} - N_{bA}\right) = 0, \quad \sum_{b=A,T,C,G} \left(N_{Cb} - N_{bC}\right) = 0, \tag{9}$$

so that only eight independent parameters should result, while the difficulties in defining a 10-dimer set of parameters from a given set of experimental data persist. In that case, complementary strand A/T and C/G pairing symmetry in a dimer, as expressed in Eq. 3, gives the conjugate NN base component relations as follows:

$$\begin{aligned} b'_x(1) &= -b_x(2);\; b'_x(2) = -b_x(1), \\ b'_y(1) &= -b_y(2);\; b'_y(2) = -b_y(1), \\ b'_z(1) &= b_z(2);\; b'_z(2) = b_z(1), \end{aligned} \tag{10}$$

where primed bases correspond to the complementary dimer and numerals correspond to the first and second nucleotides along 5′-3′ direction for each strand, that is, both order and *x,y* coordinates are inverted for the conjugate pair.

The double-strand expansion can be given as a function of a single-strand sequence taking into account the aforementioned implicit symmetries (by adding contributions from both strands to Eq. 7 taking into account Eq. 10 and then redefining the tensor set, that is, $\mathbf{E}'_{b1b2} = \mathbf{E}_{b1b2} + \mathbf{E}_{b1'b2'}$). It is clear in that case that

$$V_x = V_z = 0,\; M_{xy} = M_{yx},\; M_{xz} = -M_{zx},\; M_{yz} = -M_{zy} \tag{11}$$

correctly reducing the number of independent elementary tensor set values to 8. From Eqs. 7 and 11, the decomposition for the 10 paired NNs gives a self-consistent set of expectations obeying

$$\begin{aligned} \mathbf{E}_{TA} &= S + V_z - M_{xx} - M_{yy} + M_{zz} - 2M_{xy} - 2M_{xz} - 2M_{yz} \\ \mathbf{E}_{AT} &= S + V_z - M_{xx} - M_{yy} + M_{zz} - 2M_{xy} + 2M_{xz} + 2M_{yz} \\ \mathbf{E}_{AA-TT} &= S + V_z + M_{xx} + M_{yy} + M_{zz} + 2M_{xy} \\ \mathbf{E}_{AG-CT} &= S + M_{xx} - M_{yy} - M_{zz} - 2M_{xz} \\ \mathbf{E}_{GA-TC} &= S + M_{xx} - M_{yy} - M_{zz} + 2M_{xz} \\ \mathbf{E}_{AC-GT} &= S - M_{xx} + M_{yy} - M_{zz} - 2M_{yz} \\ \mathbf{E}_{CA-TG} &= S - M_{xx} + M_{yy} - M_{zz} + 2M_{yz} \\ \mathbf{E}_{GG-CC} &= S - V_z + M_{xx} + M_{yy} + M_{zz} - 2M_{xy} \\ \mathbf{E}_{CG} &= S - V_z - M_{xx} - M_{yy} + M_{zz} + 2M_{xy} + 2M_{xz} - 2M_{yz} \\ \mathbf{E}_{GC} &= S - V_z - M_{xx} - M_{yy} + M_{zz} + 2M_{xy} - 2M_{xz} + 2M_{yz} \end{aligned} \tag{12}$$

while the symmetrical set of eight tensor parameters can be inferred from the inverse relations

$$
\begin{aligned}
S &= \frac{1}{16}\left[2\left(\mathbf{E}_{AA-TT} + \mathbf{E}_{AG-CT} + \mathbf{E}_{GA-TC} + \mathbf{E}_{AC-GT} + \mathbf{E}_{CA-TG} + \mathbf{E}_{GG-CC}\right) + \left(\mathbf{E}_{TA} + \mathbf{E}_{AT} + \mathbf{E}_{CG} + \mathbf{E}_{GC}\right)\right] \\
V_Z &= \frac{1}{8}\left[2\left(\mathbf{E}_{AA-TT} - \mathbf{E}_{GG-CC}\right) + \left(\mathbf{E}_{TA} + \mathbf{E}_{AT} - \mathbf{E}_{CG} - \mathbf{E}_{GC}\right)\right] \\
M_{xx} &= \frac{1}{16}\left[2\left(\mathbf{E}_{AA-TT} + \mathbf{E}_{AG-CT} + \mathbf{E}_{GA-TC} - \mathbf{E}_{AC-GT} - \mathbf{E}_{CA-TG} + \mathbf{E}_{GG-CC}\right) - \left(\mathbf{E}_{TA} + \mathbf{E}_{AT} + \mathbf{E}_{CG} + \mathbf{E}_{GC}\right)\right] \\
M_{yy} &= \frac{1}{16}\left[2\left(\mathbf{E}_{AA-TT} - \mathbf{E}_{AG-CT} - \mathbf{E}_{GA-TC} + \mathbf{E}_{AC-GT} + \mathbf{E}_{CA-TG} + \mathbf{E}_{GG-CC}\right) - \left(\mathbf{E}_{TA} + \mathbf{E}_{AT} + \mathbf{E}_{CG} + \mathbf{E}_{GC}\right)\right] \\
M_{zz} &= \frac{1}{16}\left[2\left(\mathbf{E}_{AA-TT} - \mathbf{E}_{AG-CT} - \mathbf{E}_{GA-TC} - \mathbf{E}_{AC-GT} - \mathbf{E}_{CA-TG} + \mathbf{E}_{GG-CC}\right) + \left(\mathbf{E}_{TA} + \mathbf{E}_{AT} + \mathbf{E}_{CG} + \mathbf{E}_{GC}\right)\right] \\
M_{xy} &= \frac{1}{16}\left[2\left(\mathbf{E}_{AA-TT} - \mathbf{E}_{GG-CC}\right) - \left(\mathbf{E}_{TA} + \mathbf{E}_{AT} - \mathbf{E}_{CG} - \mathbf{E}_{GC}\right)\right] \\
M_{xz} &= \frac{1}{16x-8}\left(-\mathbf{E}_{TA} + \mathbf{E}_{AT} + \mathbf{E}_{CG} - \mathbf{E}_{GC}\right) = \frac{1}{4}\left(-\mathbf{E}_{AG-CT} + \mathbf{E}_{GA-TC}\right) \\
M_{yz} &= \frac{1}{16x-8}\left(-\mathbf{E}_{TA} + \mathbf{E}_{AT} - \mathbf{E}_{CG} + \mathbf{E}_{GC}\right) = \frac{1}{4}\left(-\mathbf{E}_{AC-GT} + \mathbf{E}_{CA-TG}\right)
\end{aligned} \tag{13}
$$

This decomposition enlightens the meaning of the composition-free S term as the 16-dimer ensemble mean expectation value and of V_z as the half-differential expectation between AT-containing and CG-containing dimers. Most importantly, the double determination of M_{xz} and M_{yz} values in the last two expressions in Eq. 13 should coincide for a self-consistent set of dimer values. Explicitly, self-consistency introduces links relating to composition order symmetry among dimer properties as follows:

$$
\begin{aligned}
\mathbf{E}_{AT} - \mathbf{E}_{TA} + \mathbf{E}_{CG} - \mathbf{E}_{GC} &= 2\left(\mathbf{E}_{GA-TC} - \mathbf{E}_{AG-CT}\right) \\
\mathbf{E}_{AT} - \mathbf{E}_{TA} + \mathbf{E}_{GC} - \mathbf{E}_{CG} &= 2\left(\mathbf{E}_{CA-TG} - \mathbf{E}_{AC-GT}\right).
\end{aligned} \tag{14}
$$

Note that, analogous to the composition closure relations (Eq. 9), the dimer expectation self-consistency relations (Eq. 14) may also be combined to read as follows:

$$
\begin{aligned}
\sum_{b=A,T,C,G} \left(\mathbf{E}_{Ab} - \mathbf{E}_{bA}\right) &= 0, \\
\sum_{b=A,T,C,G} \left(\mathbf{E}_{Cb} - \mathbf{E}_{bC}\right) &= 0.
\end{aligned} \tag{15}
$$

5. The modeling based on end effects

From now, we proceed to extend the irreducible model to investigate how it accommodates end effects. For the case of circular DNA, or even, for a DNA polymer, knowing the eight (polymeric) irreducible parameters (S, V_z, and the six elements of the M matrix) is sufficient for the prediction of additive physical properties. For an oligomer, additional end effects would become important and would need to be accounted for. Thus, to correctly account such effects for, consider the following duplex sequence as follows:

$$\begin{aligned} &E\, b_1 b_2 b_3 \cdots b_N\, E \\ &E b'_1 b'_2 b'_3 \cdots b'_N E, \end{aligned} \quad (16)$$

where, according to the notation introduced by Gray [10, 11], E is a pseudo-base indicating the terminations of the sequence. Pseudo-base E simply would represent one of the NNs to the end base pairs, and, under this viewpoint, it indicates interactions between the end base pairs and the surrounding solvent. Following the reasoning line suggested by Licınio and Guerra [13] and introduced in Section 3 of this review, $|A\rangle$, $|T\rangle$, $|C\rangle$, and $|G\rangle$, in Eq. 3, would correspond to the 3D part of 4D vectors with the fourth component equals to zero, and $|E\rangle$ would be a new molecular state, linearly independent with $|A\rangle$, $|T\rangle$, $|C\rangle$, and $|G\rangle$, and written as follows:

$$|E\rangle = \begin{pmatrix} 0 \\ 0 \\ 0 \\ 1 \end{pmatrix} \quad (17)$$

Then, applying Eq. 7, and, considering Eq. 11, for the duplex dimer $Eb_1 - b'_1 E$, we obtain the contribution of the end base pair b_1 / b'_1 for the thermodynamical stability of the sequence, and analog reasoning can be applied for the end base pair b_N / b'_N. Thus, for the pseudo-duplex dimer $Eb_1 - b'_1 E$,

$$\mathbf{E}(Eb_1) = A + Bx_1 + Cy_1 + Dz_{1,} \quad (18)$$

where A, B, C, and, D are parameters that determine the property under consideration. And for the pseudo-duplex dimer $b_N E - Eb'_N$, :

$$\mathbf{E}(b_N E) = A + Bx'_N + Cy'_N + Dz'_{N,} \quad (19)$$

where, in Eqs. 18 and 19, x_k is the x-component of the vector $|b_k$, and so on. According to Eqs. 18 and 19, the orientation of the end base pair would be important; for example, one A/T end base pair would not produce the same effect as one T/A end base pair. Therefore, at least in theory, it would be necessary to discriminate four-end pairings, which are listed in the following:

$$\begin{matrix} EA, & ET, & EC, & EG \\ ET, & EA, & EG, & EC. \end{matrix} \tag{20}$$

Finally, we can conclude that the four possible end base pairs in Eq. 20 can be expanded in terms of four parameters, namely *A*, *B*, *C*, and *D*. Consequently, for a duplex oligomer, the additional four parameters related to the ends should be added to the eight polymeric parameters already known, producing a total of 12 irreducible parameters, in the light of the modeling based on the end effects.

6. Results and discussion for the modeling based on the end effects

From now on, the thermodynamical property **E** will be, for us, the free energy of the duplex formation. According to the model based on the end effects, the free energy of a duplex sequence of *N* bases in the NN approximation could be calculated as the pairwise sum including end effects as a function of 12 irreducible parameters from Eqs. 12, 18, and, 19, as follows:

$$\Delta G_T = \Delta G(Eb_1) + \sum_{i=1}^{N-1} \Delta G(b_i b_{i+1}) + \Delta G(b_N E) + \Delta G_{\mathrm{sym}} \tag{21}$$

where $\Delta G_{\mathrm{sym}}=0.43\ \mathrm{kcal/mol}$ is a symmetric correction term applicable to self-complementary duplexes.

Simultaneous least-mean-square-deviation fit of this model to the 108 sequence data compiled by Allawi and SantaLucia [12] gave the values for the free energies, which are listed in Table 1. Guerra and Licinio [16] calculated irreducible parameters for the thermodynamic properties of free energy, entropy, and enthalpy but, in Table 1, only the irreducible parameters for free energy are shown. In Table 1, $\Delta G(ET)$ is the contribution for the free energy of the sequence from the T/A end base pair, and so on. Here, we prefer to use the irreducible parameters $\Delta G(EA)$, $\Delta G(ET)$, $\Delta G(EC)$, and $\Delta G(EG)$ in the place of the parameters *A*, *B*, *C*, and *D* as defined in Eq. 18 or 19. In fact, there is no loss of generality in the use of the firsts once that they are linear combinations of the seconds.

For comparison, we performed another calculation, supposing that the contributions from the ends do not depend on the orientation of the end base pairs, that is, an A/T end pair would contribute in the same way as a T/A end pair, as it is usually found in the literature [2–6]. As a result, we obtained Table 2.

Irreducible parameters for free energy	Values (kcal/mol)
$\Delta G(ET)$	0.94± 0.07
$\Delta G(EA)$	0.87± 0.07
$\Delta G(EG)$	0.82± 0.07
$\Delta G(EC)$	0.87± 0.06
S	−1.37± 0.02
V_z	0.57± 0.01
M_{xx}	0.04± 0.01
M_{yy}	−0.01± 0.01
M_{zz}	−0.05± 0.01
M_{xy}	−0.07± 0.01
M_{xz}	−0.02± 0.01
M_{yz}	−0.02± 0.01

Table 1. Irreducible Parameters for Free Energy at Standard Conditions (37 ℃ and 1 M Salt and DNA)

Irreducible parameters for free energy	Values (kcal/mol)
$\Delta G(EA,\ ET)$	0.91± 0.07
$\Delta G(EC,\ EG)$	0.84± 0.06
S	−1.37± 0.02
V_z	0.57± 0.01
M_{xx}	0.04± 0.01
M_{yy}	−0.01± 0.01
M_{zz}	−0.04± 0.01
M_{xy}	−0.07± 0.01
M_{xz}	−0.02± 0.01
M_{yz}	−0.03± 0.01

Table 2. Irreducible Parameters for Free Energy at Standard Conditions (37 ℃ and 1 M Salt and DNA)

Considering the values obtained for the irreducible parameters for free energy presented in Tables 1 and 2, some observations must be carried out:

1. Mean values and errors are essentially the same, independently of the modeling.
2. Defining the root-mean-square deviation per dimer χ [13, 16] as follows:

$$\chi = \sqrt{\sum_{i=1}^{108}\left[\Delta G_{\exp}(i) - \Delta G_{\text{theor.}}(i)\right]^2 / \sum_{i=1}^{108} N(i)}, \tag{22}$$

the free energy irreducible parameters in Tables 1 and 2 are such that they minimize χ. The quantityχ defines a global minimal deviation, between the theoretical values calculated from the irreducible parameter set for the free energies of the 108 sequences and the experimental values. In Eq. 22, $\Delta G_{\exp}(i)$ is the experimental value for the free energy for the ith sequence, $\Delta G_{\text{theor.}}$ (i) is its corresponding theoretical value, and $\sum_{i=1}^{108} N(i)$ is the total number of duplex dimers for the ensemble of 108 sequences. The value obtained forχ considering 10 (or 12) parameters is precisely the same, namely 0.14 kcal/mol per dimer [16], which also coincides with the 12-parameter model using values reported by SantaLucia for the free energies for the 10 duplex dimers [3–6]. This means that, considering only the overall data ensemble quality, there is no practical reason to prefer a model with a greater number of parameters.

3. The intrinsic errors obtained for the contributions by the ends are sensibly larger than the errors for the other irreducible parameters. In this way, in all the decomposition schemes, the contributions of the ends are not so well defined, that is, we could not differentiate its orientation (for example, we could not differentiate A/T from T/A). Thus, or the available experimental data are not still sufficiently precise or even this modeling is still inadequate to account for end effects.

4. It is also verified that the C/G or G/C end pairing is only slightly more stable than the A/T or T/A end pairing. However, the intrinsic errors in data shown in Tables 1 and 2 are considerable, allowing for portions of the ranges of possible values of the end parameters to coincide. Thus, strictly speaking, in the modeling based on end effects, there is no differentiation between the terminal base pairs.

5. The errors of the irreducible parameters for free energy were estimated in the following way: Guerra and Licinio selected 100 sets of 80 sequences chosen randomly and then calculated the mean deviation for the parameters obtained from each set [16].

As shown, end contributions are fit with large errors to experimental data, as compared to the fits of other NN or dimer contributions. Besides A/T from T/A as well as C/G from G/C, ending contributions could not be respectively differentiated. More than that, we could not distinguish between the weak and the strong terminal base pairs. However, using both the sets, one can calculate free energies for DNA oligomers at least as well as standard models considering a larger set of parameters do [3–6]. Guerra and Licinio [16] also extended their analysis and obtained equivalent sets of irreducible parameters for enthalpy and entropy. By simultaneously minimizing the deviations from melting temperatures and entropies of the chains, they obtained the most precise set, which is capable of predicting melting temperatures for DNA chains with a standard deviation of 2.2°C for sequence against a deviation of 2.5°C for previous parameters found in the literature [3–6].

In the light of our finding, the formulation based on the use of end effects, according to the NN approach, proves to be naive, even heuristic. The extra parameters (up to now, the end parameters), which must be summed to the eight (polymeric) irreducible parameters for predicting thermodynamical properties of duplex oligomers, seem not to depend on the composition of the terminal base pairs. From now, we will invoke a new hypothesis, which will be detailed later in this review. With base on this hypothesis, we will conclude that, in the light of the NN model, 10 is the number of parameters expand the free energy of any DNA oligomers: eight (polymeric) irreducible parameters for free energy, already described, plus two parameters related to the initiation of the double helix (related to two possible base pairings).

7. The modeling based on double helix initiation parameters

Equation 21 establishes how to calculate the total free energy of a sequence of length N, according to the NN model, using the methodology based on the modeling by end effects.

On the other hand, in the statistical mechanics viewpoint, the free energy of the duplex formation ΔG_T relates to the equilibrium constant K_{eq} as follows:

$$\Delta G_T = -RT \ln K_{eq}. \tag{23}$$

Whenever nucleation is the limiting process, the two-state model establishes that once the process is initiated, the helix extends to both ends of the chain [7]. The partition function or the equilibrium constant K_{eq} for the duplex formation can then be calculated as follows:

$$K_{eq} = \sigma \prod_{i=1}^{N} s_i, \tag{24}$$

where σ is the nucleation equilibrium constant and s_i is the propagation equilibrium constant, which refers to the addition of the ith base pair to the preexisting duplex. For heteropolymers, σ and s_i depend on the composition of the chain. Inserting Eq. 24 into Eq. 23, we obtain:

$$\Delta G_T = -RT \ln \sigma - RT \sum_{i=1}^{N} \ln s_i, \tag{25}$$

that is,

$$\Delta G_T = \Delta G_{\text{nuc}} - RT\sum_{i=1}^{N} \ln s_i. \tag{26}$$

Equation 26 can be conveniently rewritten as follows:

$$\Delta G_T = \Delta G_{\text{nuc}} - RT\ln s_k - RT\sum_{\substack{i=1 \\ i\neq k}}^{N} \ln s_i. \tag{27}$$

Eqs. 26 and 27 have the same signification, but when writing Eq. 27 in the form shown, we suppose that the formation of the first base pair of the duplex occurs in the kth site. Therefore, we can see that, by comparing Eq. 27 with Eq. 21, the nucleation free energy corresponds to the end effects in the NN approach, except by the term $-RT\ln s_k$, that is,

$$\Delta G_{\text{nuc}} - RT\ln s_k = -RT\ln \sigma s_k = \Delta G(Eb_1) + \Delta G(b_N E). \tag{28}$$

Quantity $\Delta G_{\text{nuc}} - RT\ln s_k$, as shown in Eq. 28, in another way corresponds the initiation free energy, ΔG_{init}, and, correspondently, σs_k is the initiation equilibrium constant associated to the formation of the first base pair of the duplex. Furthermore, to the light of the NN modeling, the initiation free energy plays the role of the end effects. Finally, the sum of the propagation free energies corresponds, also to the light of the NN model, to the sum of the dimer free energies with the following equation:

$$-RT\sum_{\substack{i=1 \\ i\neq k}}^{N} \ln s_i = \sum_{i=1}^{N-1} \Delta G(b_i b_{i+1}). \tag{29}$$

Recently, Guerra and Licinio connected to the two approaches, namely the NN and the statistical mechanics approaches, and they calculated the equilibrium constants and free energies for nucleation and propagation of a double helix in the following transition reactions [16]:

$$\begin{aligned} &\text{poly } A + \text{poly } T \rightleftharpoons \text{poly } A\cdot T \\ &\text{poly } C + \text{poly } G \rightleftharpoons \text{poly } C\cdot G. \end{aligned} \tag{30}$$

For the above homopolymers, they obtained the following nucleation free energies, at standard 1 mol concentration:

$$\begin{aligned} &\Delta G_{\text{nuc}}(\text{poly } A\cdot T) = 1.81 \text{ kcal/mol} \\ &\Delta G_{\text{nuc}}(\text{poly}\, C\cdot G) = 1.69 \text{kcal/mol}. \end{aligned} \tag{31}$$

These values were obtained using values obtained for end effects calculated from the simultaneous least-mean-square-deviations fit of the NN model to the 108-sequence data compiled by Allawi and SantaLucia [2] and listed in Tables 1 and 2, and values experimentally obtained for A/T and C/G base pairings compiled by the Frank-Kamenetskii Group [18]. Once they obtained intrinsically large errors for the end effects, the nucleation free energies for poly $A\Delta T$ and poly $C\Delta G$ homopolymers could be considered essentially similar. This result seemed strange because nucleation free energies would depend on the oligomer composition as a whole. This could indicate that end effects, as usually accounted in the NN models, could have an improper representation, having as consequence, poor fitting parameters, and an incoherent interpretation of the nucleation. Thus, the usual modeling by end effects must be seen as a didactic and heuristic approximation for DNA properties, but a better modeling needs to be discussed.

As a more appropriate modeling is a necessity, we will look for a more precise interpretation for the nucleation free energy term in the expansion of the free energy of a duplex oligomer. For this, initially, we will write the free energy for the formation of a duplex oligomer as found in some approaches in the literature [4, 6, 19]:

$$\Delta G_T = \Delta G_{\text{init}} + \sum_{i=1}^{N-1} \Delta G\left(b_i b_{i+1}\right) + \Delta G_{\text{sym}}, \tag{32}$$

where, according such references, ΔG_{init} is the "initiation" or "nucleation" free energy. Such quantity, in accordance with these referred references, is related to the difficulty of aligning the two strands and forming the first WC base pair "nucleating" the double helix which, after this step, will propagate to the ends of the chain. Specially in the work of Manyanga et al. [19], ΔG_{init} is indiscriminately called the initiation or nucleation free energy. However, the term ΔG_{init} in Eq. 32 is the initiation free energy, as can be verified by returning to the discussion that follows Eq. 28. In fact, Eq. 28 shows that nucleation free energy ΔG_{nuc} is obtained from the initiation free energy ΔG_{init} by adding a term related to the "propagation" of the WC first base pair, $-RT\ln s_k$. Therefore, it becomes clear, from now, that the terms of initiation and nucleation free energies are effectively different. It is also clear that Eq. 32 has significance if and only if ΔG_{init} is the initiation free energy. Thus, we can establish the problem: How does the term ΔG_{nuc} depend on the sequence composition? Answering to this question will help us to understand why the modeling by end effects that have been used is theoretically incorrect.

The question posed in the last paragraph will guide us throughout this section. To answer it, consider, initially, the general reaction of formation of a double helix of length N. Such duplex is formed from two separated and complementary strands S and S'. This process is the chemical reaction $S + S' \rightleftarrows S \cdot S'$. Figure 3 shows a scheme of the status of the two strands before and after the nucleation of the double helix. Before the nucleation, all the bases in each one of the two strands occupy the single strand state, and the two strands are sufficiently distant one from the other. Thereafter, during the nucleation, all the bases continue in the single strand state, but the strands are approaching one to the other via juxtaposition between the bases b_k and $b_k^{'}$ ($1 \leq k \leq N$). We suppose, with this, that the nucleation occurs in the kth site of the double chain. Finally, after the nucleation, the formation of the WC first base pair occurs, that is, the base pair $b_k / b_k^{'}$ is formed. Succeeding the nucleation event and the formation of the first base pair, we have the propagation of the double helix to both the directions, extending to the two ends of the chain, if the transition is a two-state process. As shown in Figure 3, the formation of the first WC base pair is constituted by one nucleation step followed by one propagation step. Therefore, the equilibrium constant σs_k refers to the formation of the first WC base pair, through the establishment of hydrogen bonds between the bases b_k and $b_k^{'}$. If the free energy associated to the formation of the first base pair is ΔG_{init}, then we can write the equilibrium constant σs_k as

$$\sigma s_k = \exp\left\{-\frac{\Delta G_{\text{init}}}{RT}\right\} \tag{33}$$

that is,

$$\Delta G_{\text{init}} = -RT \ln \sigma s_k = \Delta G_{\text{nuc}} - RT \ln s_k. \tag{34}$$

In order to consider the propagation of the double helix from the nucleating base pair $b_k / b_k^{'}$ and extending to both the ends, Eq. 24 can be modified for to produce:

$$K_{\text{eq}} = \left(\prod_{i=1}^{k-1} s_i^{\leftarrow}\right) \sigma s_k \left(\prod_{i=k+1}^{N} s_i^{\rightarrow}\right) \tag{35}$$

In Eq. 35, σ is the nucleation equilibrium constant, $\kappa = \sigma s_k$ is the initiation equilibrium constant (which is evidently related to the process of formation of the WC first base pair $b_k / b_k^{'}$), and $s_i^{\leftarrow}$ ($i < k$) and $s_i^{\rightarrow}$ ($i > k$) are the propagation equilibrium constants related to the propagation of the double helix, by stacking of the base pair $b_i / b_i^{'}$ on the preexistent duplex, respectively, in the 3′–5′ (downward) and 5′–3′ (upward) directions. Thus, substituting Eq. 35 into Eq. 23, we obtain

$$\Delta G_T = -RT \ln\left[\left(\prod_{i=1}^{k-1} s_i^{\leftarrow}\right) \sigma s_k \left(\prod_{i=k+1}^{N} s_i^{\rightarrow}\right)\right] = -\sum_{i=1}^{k-1} RT \ln s_i^{\leftarrow} - RT \ln \sigma s_k - \sum_{i=k+1}^{N} RT \ln s_i^{\rightarrow}. \tag{36}$$

http://dx.doi.org/10.5772/62574

As the propagation equilibrium constant depends on the local composition, we associate to the propagation equilibrium constant for the addition of the *i*th base pair, in downward direction, a value such that

$$-RT\ln s_i^{\leftarrow} = \Delta G\left(b_i b_{i+1}\right) \tag{37}$$

.

Analogously, the propagation equilibrium constant for the addition of the *i*+1th base pair, in upward direction, assumes a value such that

$$-RT\ln s_{i+1}^{\rightarrow} = \Delta G\left(b_i b_{i+1}\right) \tag{38}$$

Thus, from Eqs. 37 and 38, the propagation equilibrium constants would be, to the light of the NN approach, given by

$$s_i^{\leftarrow} = s_{i+1}^{\rightarrow} = \exp\left[-\frac{\Delta G\left(b_i b_{i+1}\right)}{RT}\right] \tag{39}$$

The first summation in Eq. 36, $-\sum_{i=1}^{k-1} RT\ln s_i^{\leftarrow}$, refers to the sum of the free energies of all the duplex dimers in downward direction related to the nucleating base pair $b_k / b_k^{'}$. In another words, such term is the total free energy related to the propagation of the double helix, starting from the nucleating base pair $b_k / b_k^{'}$ and propagating in downward direction. From Eq. 37, we have clearly that $-\sum_{i=1}^{k-1} RT\ln s_i^{\leftarrow} = \sum_{i=1}^{k-1} \Delta G(b_i b_{i+1})$. Now speaking about the second summation in Eq. 36, $-\sum_{i=k+1}^{N} RT\ln s_i^{\rightarrow}$, it refers to the free energies of all the duplex dimers in upward direction related to the base pair $b_k / b_k^{'}$, that is, it is the total free energy related to the propagation of the double helix, starting from the base pair $b_k / b_k^{'}$ and propagating in upward direction. Applying Eq. 38, we have $-\sum_{i=k+1}^{N} RT\ln s_i^{\rightarrow} = \sum_{i=k}^{N-1} \Delta G(b_i b_{i+1})$. Thus, Eq. 36 can be rewritten as follows:

$$\begin{aligned}\Delta G_T &= -RT\ln\left[\left(\prod_{i=1}^{k-1} s_i^{\leftarrow}\right)\sigma s_k\left(\prod_{i=k+1}^{N} s_i^{\rightarrow}\right)\right] \\ &= \sum_{i=1}^{k-1}\Delta G\left(b_i b_{i+1}\right) - RT\ln\sigma s_k + \sum_{i=k}^{N-1}\Delta G\left(b_i b_{i+1}\right),\end{aligned} \tag{40}$$

that is,

$$\Delta G_T = -RT\ln\left[\left(\prod_{i=1}^{k-1} s_i^{\leftarrow}\right)\sigma s_k\left(\prod_{i=k+1}^{N} s_i^{\rightarrow}\right)\right] = -RT\ln\sigma s_k + \sum_{i=1}^{N-1}\Delta G\left(b_i b_{i+1}\right), \tag{41}$$

where $\Delta G_{\text{init}} = -RT\ln\sigma s_k$ is the initiation free energy, and $\sum_{i=1}^{N-1}\Delta G(b_i b_{i+1})$ is the sum of the dimer free energies. Defining $\Delta G(Ob_k) = -RT\ln s_k$ the free energy change associated to the process of the "propagation" of the first WC base pair, we can rewrite Eq. 41 as

$$\Delta G_T = \Delta G_{\text{init}} + \sum_{i=1}^{N-1}\Delta G\left(b_i b_{i+1}\right) = \Delta G_{\text{nuc}} + \Delta G\left(Ob_k\right) + \sum_{i=1}^{N-1}\Delta G\left(b_i b_{i+1}\right). \tag{42}$$

Equation 42 shows that the free energy for the duplex formation can be written in terms of the initiation or the nucleation free energy, producing two approaches completely equivalent (the two equalities in Eq. 42). We will prefer, however, the first because it permits to obtain directly the initiation free energy for the duplex formation, as it will be shown in the next section. In addition, the nucleation free energy can be calculated from the initiation free energy, as shown in Eq. 34. Then, for applying Eq. 42, we will assume that the event of nucleation can occur by approaching the strands to each other via juxtaposition between any bases b_k and $b_k^{'}$ $(1 \le k \le N)$, with equal probability. The "nucleating" base pair, in turn, can be an A/T or C/G base pair. Thus, if the event of the formation of the first base pair can occur at any site along the double chain with the same probability, we can write the observable initiation free energy as follows:

$$\left\langle \Delta G_{\text{init}} \right\rangle = p_{A/T}\Delta G^{\circ}\left(A/T\right) + p_{C/G}\Delta G^{\circ}\left(C/G\right). \tag{43}$$

In Eq. 43, $\langle \Delta G_{\text{init}} \rangle$ is the observable initiation free energy, and $p_{A/T}$ and $p_{C/G} = 1 - p_{A/T}$ are, respectively, the probabilities with which the first base pair formed in the DNA double chain is A/T and C/G base pair. Finally, $\Delta G^{\circ}(A/T)$ and $\Delta G^{\circ}(C/G)$, are the free energy changes associated to the formation of the first base pair if it is an A/T or C/G base pair, respectively. As our approach is built on the hypothesis that the event of the formation of the first base pair can occur at any site along the chain with equal probability, the probabilities $p_{A/T}$ and $p_{C/G}$ are simply the compositions of A/T and C/G base pairs. Then, we have that $p_{A/T} = \chi_{A/T} = n_{A/T}/N$, and $p_{C/G} = \chi_{C/G} = n_{C/G}/N$, where $\chi_{X/Y}$ and $n_{X/Y}$ are, in turn, respectively, the relative occurrence number (composition) and the number of *X*/*Y* base pairs occurring along the duplex oligomer in question. Equation 34 shows how the nucleation free energy can be calculated from the initiation free energy. Therefore, the observable nucleation free energy can be written as

$$\langle \Delta G_{\text{nuc}} \rangle = \langle \Delta G_{\text{init}} \rangle + \langle RT \ln s_k \rangle. \tag{44}$$

The equilibrium constant s_k is associated to the first propagation step, that is, to the formation of the first WC base pair, which can be an A/T or C/G base pair. Invoking newly our simplifying hypothesis, which establishes that the formation of the first base pair can occur with equal probability in any site along the chain, we can write that

$$\langle RT \ln s_k \rangle = - \sum_{\{b_1 b_2\}} p_{b_1 b_2} \Delta G(b_1 b_2), \tag{45}$$

where the summation is over all the possible duplex dimers occurring along the chain, that is, b_1 and b_2 can be anyone of the four nucleotides A, T, C, or G. In Eq. 45, $p_{b_1 b_2}$ is the probability with which the base pair $b_2 / b_2^{'}$ is preceded by the base pair $b_1 / b_1^{'}$. Obviously, such probabilities are equal to the compositions of dimers along the double chain, that is, $p_{b_1 b_2} = \chi_{b_1 b_2}$, where $\chi_{b_1 b_2}$ is the composition of the duplex dimer $b_1 b_2 - b_2^{'} b_1^{'}$. Therefore,

$$\langle RT \ln s_k \rangle = \langle \Delta G(Ob_k) \rangle = - \sum_{\{b_1 b_2\}} \chi_{b_1 b_2} \Delta G(b_1 b_2). \tag{46}$$

Equation 44 can be rewritten as follows:

$$\langle \Delta G_{\text{nuc}} \rangle = -\sum_{\{b_1 b_2\}} \chi_{b_1 b_2} \Delta G(b_1 b_2) + \chi_{\frac{A}{T}} \Delta G^{\circ}(A/T) + \chi_{C/G} \Delta G^{\circ}(C/G). \tag{47}$$

From Eq. 47, it becomes clear that the nucleation free energy depends on the composition of the DNA double strand due to the presence of the terms $\chi_{A/T}$, $\chi_{C/G}$, and $\chi_{b_1 b_2}$, in the right side of the equation. According to Eq. 47, as there are 10 possible duplex dimers, $\langle \Delta G_{\text{nuc}} \rangle$ must be a function of 10 parameters: the already known eight polymeric irreducible parameters plus two parameters related to the formation of the first base pair, as defined in Eq. 43. We can simplify the approach contained in Eq. 47, discriminating the bases b_1 and b_2 only according to the weak–strong classification criteria. In this way, Eq. 47 becomes

$$\begin{aligned} \langle \Delta G_{\text{nuc}} \rangle = & - \chi_{ww} \Delta G(ww) - \chi_{ws} \Delta G(ws) - \chi_{sw} \Delta G(sw) - \\ & - \chi_{ss} \Delta G(ss) + \chi_{w} \Delta G^{\circ}(A/T) + \chi_{s} \Delta G^{\circ}(C/G). \end{aligned} \tag{48}$$

where $\Delta G(ww)$ is the mean free energy of a stack of two weak base pairs, χ_{ww} is its composition, and so on. Using Eq. 12, we obtain the following values for the mean dimer free energies:

$$\begin{aligned} \Delta G(ww) &= S + V_z + M_{zz} \\ \Delta G(ss) &= S - V_z + M_{zz} \\ \Delta G(ws) = \Delta G(sw) &= S - M_{zz}. \end{aligned} \tag{49}$$

Inserting Eq. 49 into Eq. 48, we can obtain:

$$\begin{aligned} \langle \Delta G_{\text{nuc}} \rangle = &- S - V_z(\chi_{ww} - \chi_{ss}) - M_{zz}(\chi_{ww} + \chi_{ss} - \chi_{ws} - \chi_{sw}) + \\ &+ \chi_w \Delta G^{\circ}(A/T) + \chi_s \Delta G^{\circ}(C/G) \end{aligned} \tag{50}$$

Equation 42 can be used to predict the free energy of any duplex oligomer if we know the values of all the polymeric irreducible parameters for free energy plus the free energy changes associated to the formation of the first base pair. Now, we can return to the set of 108 sequences compiled by Allawi and SantaLucia to obtain the set of eight polymeric irreducible parameters together with these two additional parameters. This will be done in the following section.

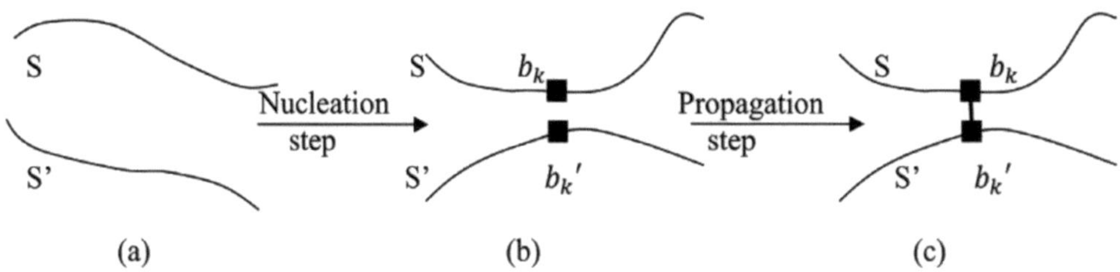

Figure 3. The formation of the first WC base pair. (a) Strands *S* and *S′* are sufficiently distant one from the other. All the bases in both the chains are in the single strand state. (b) It occurs an approximation between strands *S* and *S′*. However, all the bases are still in the single strand state. (c) It is formed the first base pair, namely the base pair $b_k / b_k^{'}$, through the establishment of H bonds between the bases b_k and $b_k^{'}$. After that, the double helix propagates in both the directions extending to the ends of the chain [17].

8. Results and discussion for the modeling based on double helix initiation parameters

The free energy for a duplex sequence of *N* bases in the NN approximation can be calculated as the pairwise sum, using the initiation free energy, as a function of the 10 parameters for free energy from Eqs. 12 and 43 as follows:

$$\left\langle \Delta G_{\text{init}} \right\rangle + \sum_{i=1}^{N-1} \Delta G\left(b_i b_{i+1}\right) + \Delta G_{\text{sym}} \tag{51}$$

Simultaneous least-mean-square-deviation fit of this model to the 108 sequence data set compiled by Allawi and SantaLucia [2] gave the values for the free energy parameters, as listed in Table 3 [17].

Irreducible parameters for free energy	Values (kcal/mol)
$\Delta G^{\circ}(A/T)$	1.7± 0.3
$\Delta G^{\circ}(C/G)$	1.8± 0.2
S	−1.38± 0.02
V_z	0.58± 0.04
M_{xx}	0.04± 0.01
M_{yy}	−0.02± 0.01
M_{zz}	−0.05± 0.01
M_{xy}	−0.07± 0.01
M_{xz}	−0.03± 0.01
M_{yz}	−0.03± 0.01

Table 3. Irreducible Parameters for Free Energy at Standard Conditions (37°C and 1 M Salt and DNA)

Given the root-mean-square deviation per dimer, as defined in Eq. 22, the parameters for free energy in Table 3 are those that minimize χ. The value obtained for χ was 0.14 kcal/mol per dimer [17], which coincides precisely with that obtained for the 12 parameter models using values reported by SantaLucia for the free energies of the 10 duplex dimers [2, 4–6]. Thus, how it happened for the modeling by end effects, from the overall data ensemble quality, there would not be practical reason to prefer a model with a greater number of parameters. The mean values and the errors of the parameters for free energy, as listed in Table 3, were estimated by Guerra in the following way [17]: he selected 1000 sets of 70 sequences chosen randomly, and then he calculated the mean and the deviation for the parameters obtained from each set. Some immediate conclusions can be done with respect to the data contained in Table 3. First, the intrinsic errors of the free energies related to the formation of the first base pairings are only a little larger than the errors of the other irreducible parameters for free energy. Second, considering only the bar of errors, the free energy changes for the initiation of a double chain through the formation of an A/T or C/G base pair are essentially similar. Thus, if it is correct the hypothesis of that the duplex formation can be initiated by the formation of a base pair at any site along the double helix with equal probability (independently of the local composition), then, the initiation free energy is essentially independent on the local composition. Finally, once we have obtained the initiation free energy parameters, as listed in Table 3, we are ready for to estimate the nucleation free energy of any duplex oligomer, using Eq. 50. Equation 50 establishes that observable nucleation free energies depend on the mean global composition of the DNA double strand and vary within a range that goes from

$$\Delta G_{\text{nuc}}^{\text{poly } A\cdot T} = 1.38 - 0.58 + 0.04 + 1.7 = 2.54 \frac{\text{kcal}}{\text{mol}},$$

for a poly $A\cdot T$ homopolymer to

$$\Delta G_{\text{nuc}}^{\text{poly } C\cdot G} = 1.38 + 0.58 + 0.04 + 1.8 = 3.80 \frac{\text{kcal}}{\text{mol}},$$

for a poly $C\cdot G$ homopolymer. Observe that the difference between these values for nucleation free energies, which is ~1.3 kcal/mol, is greater than the bar of errors estimated for nucleation free energies, which is ~0.7 kcal/mol. On the another hand, the results obtained above, for the poly $A\cdot T$ and poly $C\cdot G$ homopolymers, are in total discordance with results obtained previously using the modeling by end effects [16], as was to be expected. In fact, heteropolymers must have one value for the nucleation free energy that must be inside such interval, and it must depend on their composition. Finally, the mean observable nucleation free energy is $\Delta G_{\text{nuc}}^{\text{mean}} = \left(\Delta G_{\text{nuc}}^{\text{poly } A\cdot T} + \Delta G_{\text{nuc}}^{\text{poly } C\cdot G}\right)/2 = 3.17\ \text{kcal}/\text{mol}$. This value is only a little lower than that obtained by Manyanga et al. [19].

Comparing the results obtained for the eight polymeric irreducible parameters for free energy, as listed in Table 3 of this section, with results obtained recently using the end effects [16], and contained in Tables 1 and 2 of the Section 6 of this review, we can conclude that the irreducible parameters are not essentially affected with the alteration in the modeling. In another words, if we substitute one modeling by another, the end effects, which, obviously, do not depend essentially on the compositions of the two terminal base pairs, are substituted by the initiation free energies, which do not depend essentially on the global composition of the chain. Therefore, dimer free energies, which depend only on the irreducible parameters for free energy, also are not essentially affected.

Free energy changes associated to the formation of the second base pair are given by the following equation:

$$\left\langle \Delta G\left(b_{k\pm1} / b'_{k\pm1}\right)\right\rangle = \left\langle \Delta G\left(b_k / b'_k\right)\right\rangle + \left\langle \Delta G\left(Ob_k\right)\right\rangle \tag{52}$$

depending if the second base pair formed is located at the k+1th site or at the k−1th site of the chain. Using Eq. 43 for $\langle \Delta G(b_k / b'_k)\rangle = \langle \Delta G_{\text{init}}\rangle$, Eq. 46 for $\langle \Delta G(Ob_k)\rangle$, and also the approximations given by Eq. 49, we obtain the following:

http://dx.doi.org/10.5772/62574

$$\langle \Delta G(A/T) \rangle_{\text{base pairing}} = (0.7 \pm 0.3) \text{ kcal/mol}$$

and

$$\langle \Delta G(C/G) \rangle_{\text{base pairing}} = (0.1 \pm 0.3) \text{kcal/mol.}$$

The values listed above are just the base pairing contributions for the dimer free energies, which were encountered experimentally by the Frank-Kamenetskii Group [18]. Yakovchuk et al. obtained for the A/T and C/G base pairings, the base pairing free energies of 0.57 kcal/mol and −0.11 kcal/mol, respectively [18]. Therefore, we have obtained values that agree reasonably well with those obtained by the Frank-Kamenetskii Group. In addition, the values for the base pairing free energies are reasonably well defined because their ranges of allowable values have only an unique common intercept.

9. Conclusions

A geometrical representation of four-nucleotide sets as a tetrahedron (Eq. 3 and Fig. 2) allows for the association of the three most distinctive molecular group classifications with corresponding orthogonal cubic axis. Physical properties of nucleotide sequences may be calculated with an optimal set of tensor coefficients (Eq. 4), assuming projections within this tetrahedral representation. The coefficients are expressed in hierarchical differential form, so lower levels of approximation are explicitly embodied in the description. This includes an ensemble mean expectation from scalar coefficient S alone and a global composition approximation, as expressed through V-component contributions. The symmetrical set is shown to provide a frame for the analysis of DNA duplex free energy fully compatible with experimental data. Such a symmetrical set of coefficients allows for the translation among different decomposition frames. It also gives a proper irreducible representation for dimer properties (Eqs. 8 and 12). It solves an old indeterminacy of dimer sets by establishing self-consistency relations among the dimer coefficients (Eqs. 14 and 15).

Using the modeling based on end effects, for predicting correctly physical properties of duplex oligomers, we saw that end contributions are fit with large errors to experimental data, as compared to the fits of other NN or dimer contributions. Besides, we could not distinguish between the weak and the strong terminal base pairs. However, using both the sets constituted by two- or four-ending parameters, one calculates free energies for DNA oligomers at least as well as standard models, considering a larger set of parameters do [2, 4–6].

The modeling based on the double helix initiation parameters substitutes the end effects by the initiation parameters. The free energy changes associated to the formation of the first base pair, in the duplex formation, are fit to experimental data with errors only slightly larger than those for the NN or dimer contributions. Furthermore, we obtained that the values for the first base pairing free energies are essentially similar (because the difference between them had a value smaller than the estimated bar of errors). Thus, this could indicate an invariance of the initiation free energy with respect to the composition of the chain. Nucleation free energy, however, depends on the composition, and it can be calculated from the initiation free energy by using Eq. 34. What supports this statement is the fact of that the difference between its maximal and minimal values is larger than the error bars. The model based on the double helix initiation parameters is constructed by using the simplifying hypothesis, which establishes that the nucleation can occur at any site of the chain with equal probability, independently of the local composition. An important result, which becomes such hypothesis quite reasonable, is the fact of that the base pairing contributions for the dimer free energies seem to agree well with values experimentally obtained by the Frank-Kamenetskii Group. Finally, this modeling uses a set of 10 parameters, which is constituted by the eight polymeric irreducible parameters already known plus two parameters related to two possible base pairings (the initiation free energy parameters). With this set, one calculates free energies for DNA oligomers at least as well as standard models considering a larger set of parameters do.

Author details

João C. O. Guerra*

Address all correspondence to: jcog@infis.ufu.br

Instituto de Física, Universidade Federal de Uberlândia, Uberlândia, MG, Brazil

References

[1] Gray D M, Tinoco I, Jr. A new approach to study of sequence-dependent properties of nucleotides. Biopolymers. 1970; 9: 223–244. DOI:10.1002/bip.1970.360090207

[2] Allawi H T, SantaLucia J, Jr. Thermodynamics and NMR of internal G T mismatches in DNA. Biochemistry. 1997; 36: 10581–10594. DOI:10.1021/bi962590c

[3] Breslauer K J, Frank R, Blocker H, Marky L A. Predicting DNA duplex stability from the base sequences. Proc Natl Acad Sci USA. 1986; 83: 3746–3750.

[4] SantaLucia J, Jr. A unified view of polymer, dumbbell, and oligonucleotide DNA nearest-neighbor thermodynamics. Proc Natl Acad Sci USA. 1998; 95: 1460–1465.

[5] SantaLucia J, Jr., Hicks D. The thermodynamics of DNA structural motifs. Annu Rev Biophys Biomol Struct. 2004; 33: 415–440. DOI:10.1146/annurev.biophys.32.110601.141800

[6] SantaLucia J, Jr., Allawi H T, Seneviratne P A. Improved nearest-neighbor parameters for predicting DNA duplex stability. Biochemistry. 1986; 35: 3555–3562. DOI: 10.1021/bi951907q

[7] Cantor C R, Schimmel P R. Biophysical Chemistry Part III: The Behavior of Biological Macromolecules. San Francisco: Freeman; 1980.

[8] Vologodskii A V, Amirikyan B R, Lyubchenko Y L, Frank-Kamenetskii M D. Allowance for heterogeneous stacking in the DNA helix-coil transition theory. J Biomol Struct Dyn. 1984; 2: 131–148. DOI:10.1080/07391102.1984.10507552

[9] Goldstein R F, Benight A S. How many numbers are required to specify sequence-dependent properties of polynucleotides? Biopolymers. 1992; 32: 1679–1693. DOI: 10.1002/bip.360321210

[10] Gray D M. Derivation of nearest-neighbor properties from data on nucleic acid oligomers. II. Thermodynamic parameters of DNA RNA hybrids and DNA duplexes. Biopolymers. 1997; 42: 795–810. DOI:10.1002/(SICI)1097-0282(199712)42:7<795::AID-BIP5>3.0.CO;2-O

[11] Gray D M. Derivation of nearest-neighbor properties from data on nucleic acid oligomers. I. Simple sets of independent sequences and the influence of absent nearest neighbors. Biopolymers. 1997; 42: 783–793. DOI:10.1002/(SICI)1097-0282(199712)42:7<783::AID-BIP4>3.0.CO;2-P

[12] Doktycz M J, Goldstein R F, Paner T M, Gallo F J, Benight A S. Studies of DNA dumbbells. I. Melting curves of 17 DNA dumbbells with different duplex stem sequences linked by T4 endloops: evaluation of the nearest-neighbor stacking interactions in DNA. Biopolymers. 1992; 32: 849–864. DOI:10.1002/bip.360320712

[13] Licinio P, Guerra J C O. Irreducible representation for nucleotide sequence physical properties and self-consistency of nearest-neighbor dimer sets. Biophys J. 2007; 92: 2000–2006. DOI:10.1529/biophysj.106.095059

[14] Licinio P, Caligiorne R B. Inference of phylogenetic distances from DNA-walk divergences. Phys A Stat Theor Phys. 2004; 341: 471–481. DOI:10.1016/j.physa.2004.03.098

[15] Xia T, SantaLucia J, Jr., Burkard M E, Kierzek R, Schroeder S J, Jiao X, Cox C, Turner D H. Thermodynamic parameters for an expanded nearest-neighbor model for formation of RNA duplexes with Watson–Crick base pairs. Biochemistry. 1998; 37: 14719–14735. DOI:10.1021/bi9809425

[16] Guerra J C O, Licinio P. Terminal contributions for duplex oligonucleotide thermodynamic properties in the context of nearest neighbor models. Biopolymers. 2011; 95: 194–201. DOI:10.1002/bip.21560

[17] Guerra J C O. Calculation of nucleation free energy for duplex oligomers in the context of nearest neighbor models. Biopolymers. 2013; 99: 538–547. DOI:10.1002/bip.22214

[18] Yakovchuk P, Protozanova E, Frank-Kamenetskii M D. Base-stacking and base-pairing contributions into thermal stability of the DNA double helix. Nucleic Acids Res. 2006; 34: 564–574. DOI:10.1093/nar/gkj454

[19] Manyanga F, Horne M T, Brewood G P, Fish D J, Dickman R, Benight A S. Origins of the "nucleation" free energy in the hybridization thermodynamics of short duplex DNA. J Phys Chem B. 2009; 113: 2556–2563. DOI:10.1021/jp809541m